Graded examples in mathematics

General Arithmetic

M. R. Heylings M.A., M.Sc.

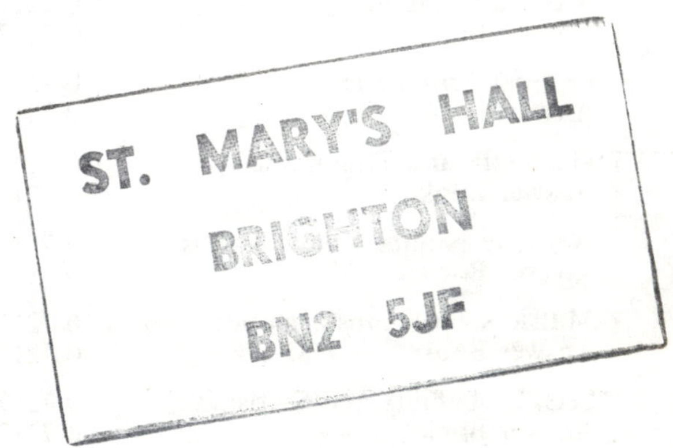

Schofield & Sims Limited   Huddersfield

0 7217 2329 2

First printed 1983
Revised and reprinted 1984
Reprinted 1986

The series **Graded examples in mathematics**
comprises:

Fractions and Decimals	0 7217 2323 3
Answer Book	0 7217 2324 1
Algebra	0 7217 2325 x
Answer Book	0 7217 2326 8
Area and Volume	0 7217 2327 6
Answer Book	0 7217 2328 4
General Arithmetic	0 7217 2329 2
Answer Book	0 7217 2330 6
Geometry and Trigonometry	0 7217 2331 4
Answer Book	0 7217 2332 2
Negative Numbers and Graphs	0 7217 2333 0
Answer Book	0 7217 2334 9
Matrices and Transformations	0 7217 2335 7
Answer Book	0 7217 2336 5
Sets, Probability and Statistics	0 7217 2337 3
Answer Book	0 7217 2338 1

In preparation:

Revision of Topics	0 7217 2339 x
Answer Book	0 7217 2340 3

Designed by Peter Sinclair (Design and Print) Ltd, Wetherby
Printed in England by Pindar Print Limited, Scarborough, North Yorkshire

Author's Note

This series has been written and produced in the form of eight topic books, each offering a wealth of graded examples for pupils in the 11–16 age range; plus a further book of revision examples for fifth formers.

There are no teaching points in the series. The intention is to meet the often heard request from teachers for a wide choice of graded examples to support their own class teaching. The contents are clearly labelled for easy use in conjunction with an existing course book; but the books can also be used as the chief source of material, in which case the restrictions imposed by the traditional type of mathematics course book are removed and the teacher is free to organise year-by-year courses to suit the school. Used in this way, the topic-book approach offers an unusual and useful continuity of work for the class-room, for homework or for revision purposes.

The material has been tested over many years in classes ranging from mixed ability 11-year-olds to fifth formers taking public examinations. Some sections are useful for pupils of above average ability while other sections suit the needs of the less able, though it is for the middle range of ability that the series is primarily intended.

Contents

Arithmetic in Society

Symbols

$=$	is equal to
$\neq$	is not equal to
$\simeq$	is approximately equal to
$<$	is less than
$\leqslant$	is less than or equal to
$\nless$	is not less than
$>$	is greater than
$\geqslant$	is greater than or equal to
$\ngtr$	is not greater than
$\Rightarrow$	implies
$\Leftarrow$	is implied by
$\rightarrow$	maps onto
$\in$	is a member of
$\notin$	is not a member of
$\subset$	is a subset of
$\not\subset$	is not a subset of
$\cap$	intersection (or overlap)
$\cup$	union
A'	the complement (or outside) of set A
$\mathscr{E}$	The Universal set
$\varnothing$ or $\{\ \}$	the empty set
(x, y)	the co-ordinates of a point
$\begin{pmatrix} x \\ y \end{pmatrix}$	the components of a vector

The Greek alphabet

A	α	alpha
B	β	beta
Γ	γ	gamma
Δ	δ	delta
E	ε	epsilon
Z	ζ	zeta
H	η	eta
Θ	θ	theta
I	ι	iota
K	κ	kappa
Λ	λ	lambda
M	μ	mu
N	ν	nu
Ξ	ξ	xi
O	o	omicron
Π	π	pi
P	ρ	rho
Σ	σ, ς	sigma
T	τ	tau
Y	υ	upsilon
Φ	ϕ, φ	phi
X	χ	chi
Ψ	ψ	psi
Ω	ω	omega

Base Arithmetic

Introduction

Part 1 Base three

Each diagram below shows a number of dots grouped in threes and put into boxes.
Three boxes are then put into a ring.

For each diagram, count (i) how many dots there are altogether
                (ii) how many rings there are
                (iii) how many loose boxes are not ringed
                (iv) how many loose dots are not boxed.

Copy the sentences and fill in the gaps as shown in this example.

Example

17 dots give .1. .2. .2. three .

1   … dots give … … … three .

2   … dots give … … … three .

3   … dots give … … … three .

4   … dots give … … … three .

5   … dots give … … … three .

6   … dots give … … … three .

Introduction

Part 2 Base four

Each diagram below shows a number of dots grouped into boxes with four dots per box. Four boxes are then put into a ring.

Copy and complete these sentences.

1 … dots give … … … four.

2 … dots give … … … four.

3 … dots give … … … four.

4 … dots give … … … four.

5 … dots give … … … four.

6 … dots give … … … four.

Introduction

Part 3 Base five

These diagrams show dots grouped in fives.
For each diagram, copy and complete this sentence.

… dots give … … … five ·

Introduction

Part 4 Different bases

These diagrams have the dots grouped in different ways.
For each diagram, copy and complete this sentence.

$$\ldots \text{ dots give } \ldots \ldots \ldots \, n.$$

Replace the n by the base used.

1 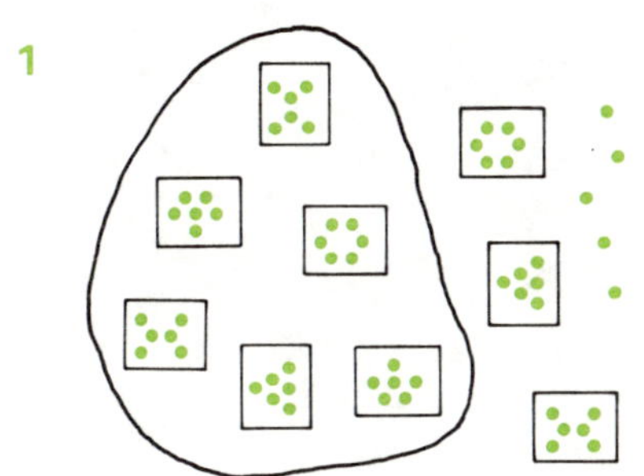

5 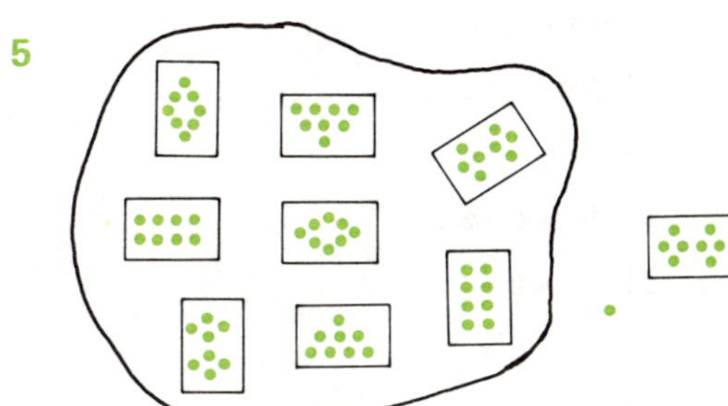

2 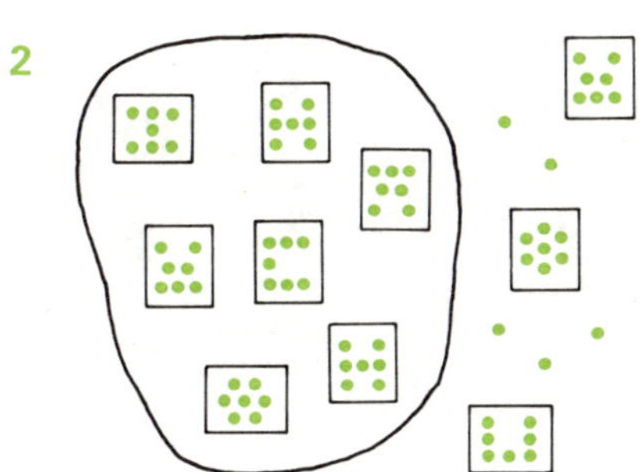

6 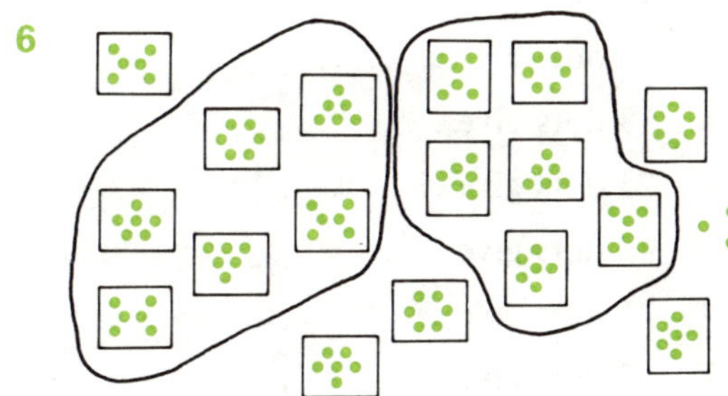

3 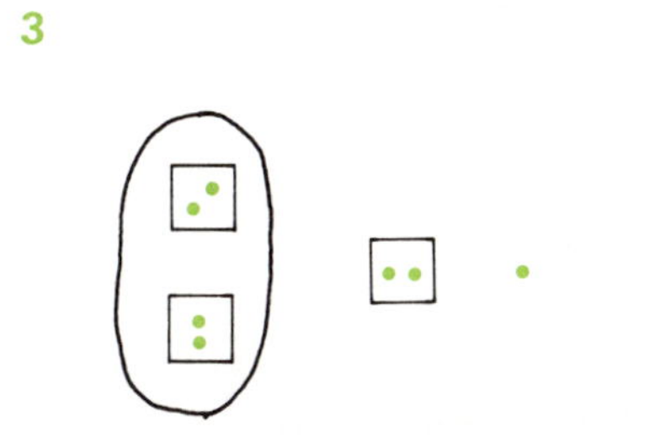

7 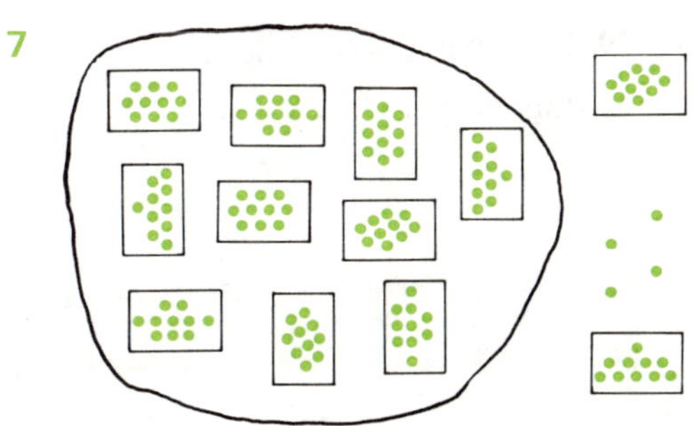

4 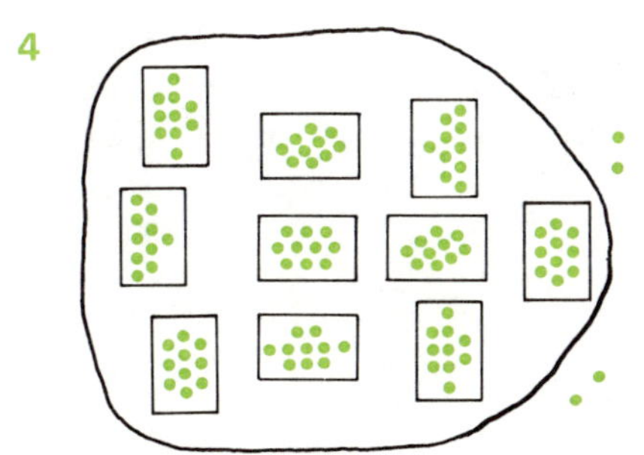

8 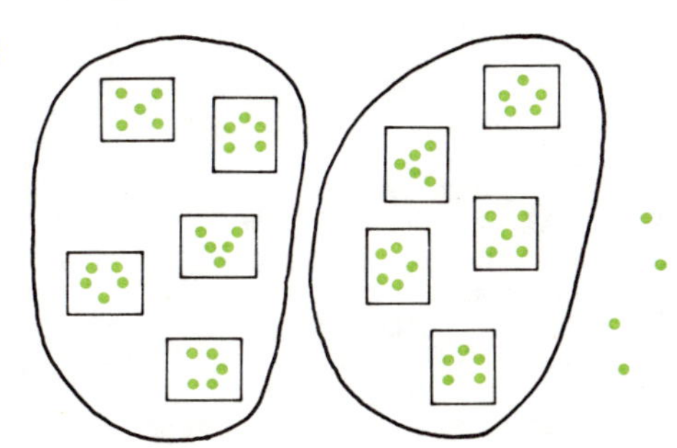

Introduction

Draw the number of dots given in each of these statements. Group them in the given base using boxes and rings.

Then complete this statement

$$\ldots \text{dots give} \ldots \ldots \ldots _n.$$

where n is replaced by the base used.

9   23 dots give … … … five ·

10   32 dots give … … … five ·

11   19 dots give … … … five ·

12   19 dots give … … … four ·

13   25 dots give … … … four ·

14   21 dots give … … … four ·

15   13 dots give … … … three ·

16   10 dots give … … … three ·

17   26 dots give … … … three ·

18   18 dots give … … … seven ·

19   23 dots give … … … ten ·

20   5 dots give … … … two ·

21   31 dots give … … … four ·

22   32 dots give … … … twelve ·

23   19 dots give … … … ten ·

24   29 dots give … … … eleven ·

25   45 dots give … … … twelve ·

Interchanging bases

Part 1 Place values

1 Base ten

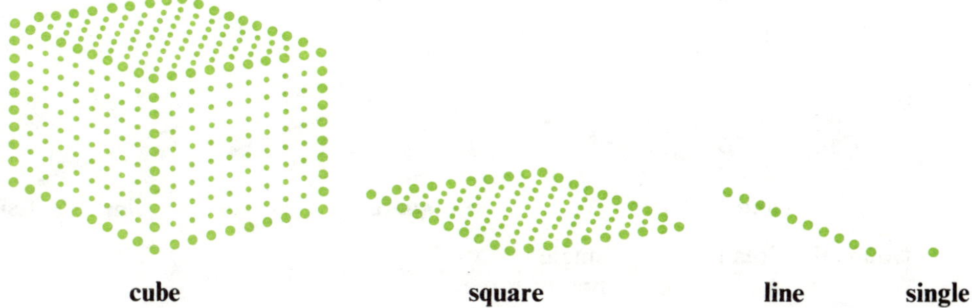

Count the dots in
  a   a single
  b   a line of ten
  c   a square of ten
  d   a cube of ten.

Write the answers in brackets as shown below. Follow the pattern of the answers to fill the two gaps. Do not write in this book, use an exercise book.

These numbers are called **place values**.

Base ten   (...)   (...)   (1000)   (100)   (10)   (.1.)

2 Base two

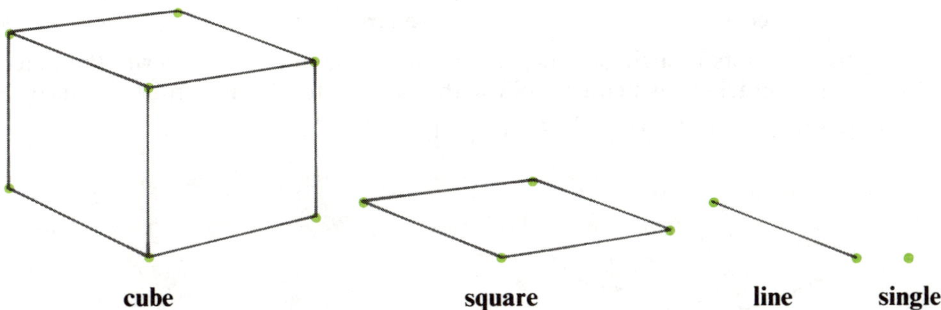

Count the dots in
  a   a single
  b   a line of two
  c   a square of two
  d   a cube of two.

Write the answers in brackets as shown below. Follow the pattern of the answers to fill the two extra place values.

Base two   (...)   (...)   (...)   (.4.)   (.2.)   (.1.)

Interchanging bases

3 Base three

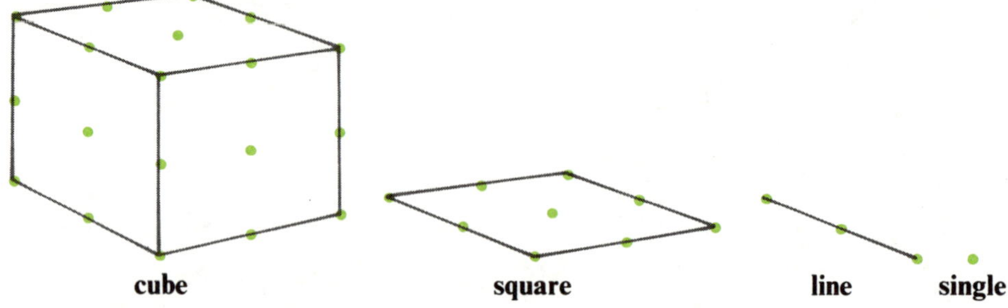

Count the dots in
a a single
b a line of three
c a square of three
d a cube of three.

Write the answers in brackets as shown below. Follow the pattern of the answers to fill the two extra place values.

Base three   (...) (...) (...) (...) (.3.) (.1.)

4 Base four

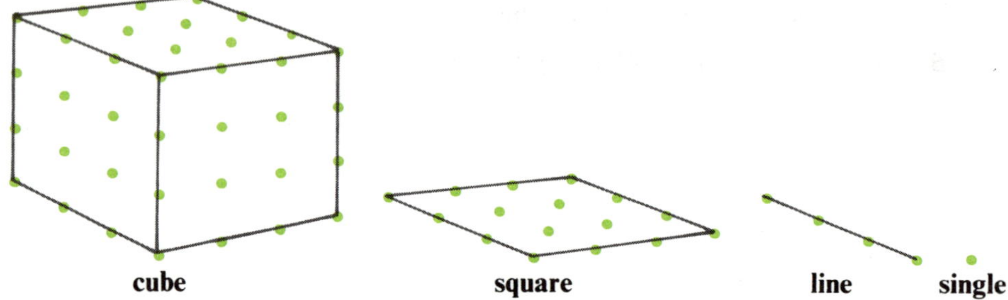

Count the dots in a single, line, square and cube of four, and write the place values in brackets as before. Follow the pattern to fill the extra two place values.

Base four   (...) (...) (...) (...) (.4.) (.1.)

5 Base five

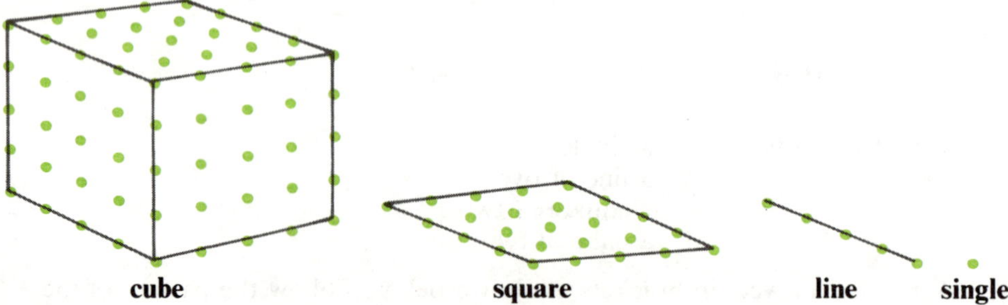

Copy and complete these brackets for the place values of base five.

Base five   (...) (...) (...) (...) (.5.) (.1.)

Interchanging bases

6 Without drawing any diagrams, copy these brackets and fill the gaps with place values for the bases given.

Base six   (…) (…) (…) (…) (.6.) (.1.)

Base seven   (…) (…) (…) (…) (.7.) (.1.)

Base eight   (…) (…) (…) (…) (.8.) (.1.)

Base nine   (…) (…) (…) (…) (…) (.1.)

Base eleven (…) (…) (…) (…) (…) (.1.)

Base twelve (…) (…) (…) (…) (…) (…)

Part 2 Changing into base ten

1 Change these base four numbers into base ten (or denary) numbers, by writing the place values as shown in this example.

Example

$$1321_{four} = \begin{matrix}(64) & (16) & (4) & (1)\\ 1 & 3 & 2 & 1\end{matrix} = 64 + 48 + 8 + 1 = \begin{matrix}64\\48\\8\\+\;1\\\hline =\;121_{ten}\end{matrix}$$

a	12	b	21	c	23	d	101
e	110	f	120	g	131	h	133
i	1011	j	1022	k	1103	l	1201
m	1212	n	1233	o	1333	p	2001
q	2131	r	2203	s	2310	t	2333
u	3001	v	3022	w	3103	x	3333
y	10012	z	11032				

2 Change these base five numbers into base ten (or denary) numbers.

a	14	b	23	c	42	d	123
e	133	f	141	g	202	h	214
i	230	j	241	k	323	l	332
m	414	n	1421	o	1304	p	1444
q	2003	r	2132	s	2401	t	11230
u	11304	v	12041	w	23124	x	33333

3 Change these base eight numbers into base ten (or denary) numbers.

a	15	b	26	c	53	d	67
e	124	f	155	g	162	h	237
i	270	j	326	k	504	l	557
m	1236	n	2504	o	3476		

Interchanging bases

4 Change these numbers with different bases into base ten numbers.

a	32_{six}	b	32_{seven}	c	143_{five}
d	143_{nine}	e	204_{six}	f	204_{eight}
g	312_{four}	h	148_{eleven}	i	402_{five}
j	1213_{four}	k	1101_{two}	l	2021_{three}
m	4003_{five}	n	1011_{two}	o	3232_{four}
p	567_{eight}	q	189_{twelve}	r	176_{eleven}
s	11011_{two}	t	10221_{three}	u	4303_{five}
v	1635_{seven}	w	20331_{four}	x	111010_{two}
y	101110_{two}	z	112202_{three}		

Bases greater than ten

Difficulty can arise with bases greater than ten, when, for example, 10_{twelve} stands for one group of twelve, not a group of ten.

The following letters can be used when just *one* symbol is needed to stand for a number greater than ten.

A = ten B = eleven C = twelve D = thirteen
E = fourteen F = fifteen etc.

5 Change these base twelve numbers into base ten.

a	1A	b	2B	c	4A	d	6B
e	A2	f	A4	g	B5	h	A9
i	AB	j	B0	k	12B	l	1A7
m	10A	n	2A6	o	25B		

6 Change these numbers with different bases into base ten.

a	$1C_{sixteen}$	b	$2D_{fourteen}$	c	$3E_{fifteen}$
d	$A5_{twelve}$	e	$A5_{sixteen}$	f	$C0_{thirteen}$
g	$C7_{thirteen}$	h	$16B_{twelve}$	i	$10F_{sixteen}$
j	$12E_{sixteen}$	k	$2A7_{fifteen}$	l	$25D_{fifteen}$
m	$34A_{eleven}$	n	$4A9_{eleven}$	o	$2FF_{sixteen}$

Harder problems

7 Each asterisk * represents a missing digit. Copy and complete.

a	$1*2_{four} = 30_{ten}$	b	$12*_{four} = 26_{ten}$
c	$21*_{three} = 23_{ten}$	d	$2*2_{three} = 20_{ten}$
e	$14*_{five} = 48_{ten}$	f	$3*2_{five} = 82_{ten}$
g	$101*1_{two} = 23_{ten}$	h	$1A*_{twelve} = 271_{ten}$
i	$12**_{four} = 100_{ten}$	j	$2*11_{four} = 13*_{ten}$

8 Find the base n in which each of these are counted.

a	$5_{ten} = 11_n$	b	$5_{ten} = 10_n$	c	$5_{ten} = 12_n$
d	$5_{ten} = 101_n$	e	$14_{ten} = 24_n$	f	$17_{ten} = 21_n$
g	$25_{ten} = 221_n$	h	$34_{ten} = 202_n$	i	$46_{ten} = 3A_n$
j	$100_{ten} = 91_n$	k	$58_{ten} = 213_n$	l	$31_{ten} = 1011_n$

Interchanging bases

Part 3 Changing from base ten

First method   Write the place values until the given number is exceeded. Fill the columns, starting with the highest possible place value. For example, to change 44_{ten} into base 3, write down (81) (27) (9) (3) (1).

Now 44 will give one 27 with remainder 17
and 17 will give one 9 with remainder 8
and 8 will give two 3s with remainder 2
and 2 will give two 1s.

$$\begin{array}{ccccc} (81) & (27) & (9) & (3) & (1) \\ \end{array}$$

So $44_{\text{ten}} = \quad 1 \quad 1 \quad 2 \quad 2 = 1122_{\text{three}}$.

Second method   Divide the given number repeatedly by the base required, and write down the remainder at each stage. Then write the remainders in reverse order.

$$
\begin{array}{r|l}
3 & 44 \\
\hline
3 & 14 \ r2 \\
\hline
3 & 4 \ r2 \\
\hline
3 & 1 \ r1 \\
\hline
 & 0 \ r1 \\
\end{array}
$$

Thus $44_{\text{ten}} = 1122_{\text{three}}$.

1   Change these base ten (denary) numbers into base four.

a	25	b	23	c	34	d	39
e	29	f	31	g	50	h	58
i	75	j	80	k	94	l	100
m	121	n	143	o	167		

2   Change these base ten (denary) numbers into base eight (octal).

a	19	b	41	c	38	d	50
e	63	f	65	g	75	h	94
i	151	j	175	k	586	l	1380

3   Change these base ten numbers into base two (binary).

a	13	b	11	c	29	d	22
e	39	f	58	g	63	h	65
i	84	j	100	k	127	l	191

4   Change these base ten numbers into the bases given.

a	16 into base three	b	59 into base five
c	70 into base six	d	131 into base nine
e	188 into base twelve	f	44 into base three
g	49 into base three	h	164 into base five
i	193 into base five	j	358 into base five
k	231 into base six	l	284 into base six
m	541 into base six	n	853 into base nine
o	939 into base nine	p	30 into base two
q	133 into base three	r	211 into base three
s	415 into base four	t	699 into base two

Interchanging bases

5 Change these base ten (denary) numbers into base twelve (duodenary). With base twelve use A for ten and B for eleven.

a	22	b	23	c	34	d	58
e	71	f	107	g	90	h	124
i	131	j	178	k	213	l	410

6 For bases greater than twelve use C for twelve, D for thirteen, E for fourteen, F for fifteen etc.

Change these base ten numbers into the bases given.

a   23 into base thirteen   b   38 into base thirteen

c   24 into base sixteen   d   31 into base sixteen

e   42 into base sixteen   f   37 into base fourteen

g   54 into base fourteen   h   154 into base fifteen

i   190 into base fifteen   j   296 into base fifteen

k   169 into base sixteen   l   440 into base sixteen.

Part 4 A mixture

1 Change these numbers into base ten.

a	135_{six}	b	241_{five}	c	1022_{three}
d	1101_{two}	e	1314_{five}	f	356_{eight}
g	748_{nine}	h	259_{twelve}	i	$1A7_{twelve}$

2 Change these base ten numbers into the bases given.

a	37_{ten} into base five	b	59_{ten} into base five
c	25_{ten} into base three	d	50_{ten} into base three
e	14_{ten} into base two	f	28_{ten} into base two
g	174_{ten} into base twelve	h	184_{ten} into base twelve
i	159_{ten} into base four		

3 Copy and complete this table.

	Base two	Base three	Base ten
a			1
b			2
c			3
d	100		
e	101		
f	110		
g		21	
h		22	
i		100	
j			10

Interchanging bases

4 Change each of these numbers into base ten, then into the base given.

a 1101_{three} into base two	**b** 234_{five} into base three
c 110101_{two} into base eight	**d** 101111_{two} into base eight
e 1231_{four} into base twelve	**f** 253_{six} into base five
g 134_{eight} into base two	**h** 2202_{three} into base four
i 3102_{four} into base nine	**j** 111010111_{two} into base sixteen
k $138_{sixteen}$ into base two	**l** $1A7_{twelve}$ into base sixteen

5 Copy and complete this table.

	a	b	c	d	e	f	g	h
Base ten	12	27						
Base four			32	113	202			
Base six						52	124	140

6 Change each of these into base ten to do the working, then change back into the base given for the answer.

a $12_{three} \times 21_{three}$	**b** $101_{two} \times 111_{two}$
c $123_{four} \times 11_{four}$	**d** $144_{five} \times 13_{five}$
e $212_{three} \times 121_{three}$	**f** $202_{three} \div 12_{three}$
g $101101_{two} \div 1001_{two}$	**h** $1430_{five} \div 11_{five}$
i $180_{twelve} \div 13_{twelve}$	**j** $1311_{four} \div 21_{four}$

k $\dfrac{1211_{three} \times 110_{three}}{210_{three}}$

l $\dfrac{13A_{twelve} \times 26_{twelve}}{21_{twelve}}$

7 A couple make monthly repayments on their house mortgage, the first payment being in January. They count the number of repayments they make in base twelve; so January is Month 1, February is Month 2 and so on. What is the name of the month which they count as Month 134?

8 A new type of calendar counts the days of the year in base seven, so that Monday the first of January is counted as Day 1, Tuesday the second of January as Day 2. On which day of the week will Day 153 fall?

9 Eggs are sold in cartons each containing six eggs. If I buy 132_{six} eggs, how many cartons have I bought and how many loose eggs (not in a carton) will I have?

10 A machine automatically stamps out the letters of the alphabet in order, starting with the letter A. If it has stamped out $107_{twenty\ six}$ letters, which letter will it stamp out next?

11 A shop assistant is stacking tins of beans on a shelf, taking them from boxes holding eight tins. As she stacks them she counts the tins in base eight. After stacking 215_{eight} tins, she is part way through a box. How many tins are still left in this box?

12 Correct the mistake in each of the following statements.

a $142_{three} = 23_{ten}$	**b** $261_{five} = 81_{ten}$
c $250_{six} = A2_{ten}$	**d** $1112_{two} = 51_{three}$

Addition and subtraction of different bases

Part 1 Base four

Work these.

1	102 + 122	**4**	213 + 103	**7**	131 + 23	**10**	133 + 303
2	131 + 112	**5**	132 + 102	**8**	201 + 331		
3	223 + 102	**6**	213 + 22	**9**	312 + 302		

11 221 + 203
12 1301 + 213
13 2033 + 201
14 1211 + 1003 + 1011
15 2021 + 1103 + 120

16 1303 + 1131 + 1100
17 333 + 1
18 333 + 33 + 3
19 223 + 112 + 233
20 123 + 333 + 321

Part 2 Base five

1	213 + 122	**4**	233 + 104	**7**	203 + 143	**10**	143 + 24
2	121 + 232	**5**	141 + 243	**8**	324 + 34		
3	304 + 122	**6**	134 + 212	**9**	204 + 43		

11 413 + 213
12 204 + 332
13 1420 + 233
14 2041 + 1104
15 1132 + 1423 + 244

16 2023 + 414 + 1104
17 444 + 1
18 3210 + 444 + 1234
19 4321 + 2222 + 1234
20 4444 + 444 + 44 + 4

Part 3 Base eight

1	534 + 125	**3**	155 + 426	**5**	7405 + 614
2	371 + 264	**4**	3064 + 2136		

6 5367 + 711
7 2015 + 663
8 421 + 376 + 205
9 374 + 106 + 525

10 1244 + 616 + 106
11 3465 + 1035 + 417
12 1771 + 456 + 2363

Addition and subtraction of different bases

Part 4 Different bases

1 312_{six} $+ 215_{six}$

4 784_{twelve} $+ 182_{twelve}$

7 1001_{two} $+ 101_{two}$

10 135_{ten} 1406_{ten} $+ 327_{ten}$

2 102_{three} $+ 12_{three}$

5 1433_{seven} $+ 154_{seven}$

8 213_{four} $+ 23_{four}$

3 618_{nine} $+ 173_{nine}$

6 2205_{eight} $+ 634_{eight}$

9 1024_{six} 315_{six} $+ 2124_{six}$

11 $123 + 1012 + 402$ (base five)
12 $346 + 1214 + 706$ (base eight)
13 $43 + 124 + 51$ (base six)
14 $463 + 217 + 45$ (base twelve)
15 $306 + 41 + 182$ (base nine)
16 $202 + 12 + 112$ (base three)
17 $697 + 42 + 228$ (base sixteen)
18 $314 + 43 + 233$ (base five)
19 $1011 + 101 + 100$ (base two)
20 $532 + 144 + 51$ (base six)

Each asterisk * represents a missing digit. Copy and complete.

21 $3*5_{six}$ $+ 14*_{six}$ $\overline{ *11_{six}}$

23 $41*_{nine}$ $+ *37_{nine}$ $\overline{ 7*1_{nine}}$

22 $*64_{eight}$ $+ 2*6_{eight}$ $\overline{ 52*_{eight}}$

24 $2**_{three}$ $+ *12_{three}$ $\overline{ *120_{three}}$

Part 5 Bases greater than ten

Remember that A = ten, B = eleven, C = twelve, D = thirteen, E = fourteen, F = fifteen etc.

1 356_{twelve} $+ 547_{twelve}$

3 $719_{sixteen}$ $+ 537_{sixteen}$

5 $7A4_{fifteen}$ $+ 168_{fifteen}$

7 $1D3_{twenty}$ $+ 52A_{twenty}$

2 269_{twelve} $+ 845_{twelve}$

4 $31A_{sixteen}$ $+ A27_{sixteen}$

6 $5A_{fifteen}$ $+ 6B2_{fifteen}$

8 $6B_{twelve}$ $+ 736_{twelve}$

9 $3A + 125 + 6B1$ (base twelve)
10 $16F + 32 + 2A4$ (base twenty)
11 $50C + 27 + 916$ (base fourteen)
12 $125 + 30A + 57$ (base eleven)

13 $1AB + 47 + 20C$ (base sixteen)
14 $42C + 8F2 + 27$ (base twenty)
15 $13AC + A47 + 18B$ (base fifteen)

Part 6 Base four

1 231 $- 20$

4 113 $- 32$

7 223 $- 132$

10 311 $- 22$

2 223 $- 12$

5 321 $- 32$

8 311 $- 102$

11 101 $- 31$

3 122 $- 31$

6 212 $- 33$

9 320 $- 111$

12 302 $- 111$

Addition and subtraction of different bases

13 1331 − 1212
14 3012 − 1123
15 2130 − 212
16 200 − 13

17 300 − 22
18 301 − 123
19 2001 − 213
20 3002 − 1203

Part 7 Base five

1
$$234$$
$$- 124$$

4
$$413$$
$$- 132$$

7
$$321$$
$$- 33$$

10
$$403$$
$$- 14$$

2
$$313$$
$$- 141$$

5
$$112$$
$$- 104$$

8
$$213$$
$$- 24$$

11
$$300$$
$$- 123$$

3
$$213$$
$$- 104$$

6
$$132$$
$$- 14$$

9
$$201$$
$$- 113$$

12
$$1000$$
$$- 444$$

13 3012 − 1103
14 4031 − 1212
15 3102 − 2141
16 3021 − 423

17 100 − 1
18 202 − 114
19 1000 − 1
20 3002 − 2144

Part 8 Base eight

1
$$643$$
$$- 212$$

3
$$662$$
$$- 471$$

5
$$724$$
$$- 307$$

7
$$514$$
$$- 145$$

2
$$471$$
$$- 126$$

4
$$441$$
$$- 126$$

6
$$632$$
$$- 277$$

8
$$603$$
$$- 24$$

9 3264 − 507
10 2314 − 675

11 302 − 144
12 2004 − 35

Part 9 Different bases

1
$$342_{six}$$
$$- 151_{six}$$

4
$$321_{twelve}$$
$$- 198_{twelve}$$

7
$$441_{six}$$
$$- 53_{six}$$

10
$$1203_{four}$$
$$- 312_{four}$$

2
$$374_{eight}$$
$$- 126_{eight}$$

5
$$245_{ten}$$
$$- 184_{ten}$$

8
$$203_{five}$$
$$- 44_{five}$$

3
$$1035_{seven}$$
$$- 316_{seven}$$

6
$$322_{six}$$
$$- 45_{six}$$

9
$$304_{six}$$
$$- 25_{six}$$

11 1234 − 125 (base seven)
12 568 − 482 (base nine)
13 1010 − 101 (base two)
14 336 − 189 (base twelve)
15 5042 − 361 (base seven)

16 201 − 89 (base sixteen)
17 4200 − 142 (base five)
18 6000 − 1 (base seven)
19 3340 − 1341 (base seven)
20 321002 − 120303 (base four)

Addition and subtraction of different bases

Each asterisk * represents a missing digit. Copy and complete.

21
$$
\begin{array}{r}
***_{four} \\
- \ 231_{four} \\
\hline
22_{four}
\end{array}
$$

22
$$
\begin{array}{r}
3*2_{seven} \\
- \ *1*_{seven} \\
\hline
165_{seven}
\end{array}
$$

23
$$
\begin{array}{r}
7*3_{eight} \\
- \ *2*_{eight} \\
\hline
222_{eight}
\end{array}
$$

24
$$
\begin{array}{r}
3*4*5*_{six} \\
- \ *1*2*3_{six} \\
\hline
123321_{six}
\end{array}
$$

Part 10 Bases greater than ten

Remember that A = ten, B = eleven, C = twelve, D = thirteen, E = fourteen, F = fifteen etc.

1
$$
\begin{array}{r}
342_{twelve} \\
- \ 124_{twelve} \\
\hline
\end{array}
$$

5
$$
\begin{array}{r}
8A3_{fourteen} \\
- \ 5\,29_{fourteen} \\
\hline
\end{array}
$$

9
$$
\begin{array}{r}
2677_{sixteen} \\
- \ 829_{sixteen} \\
\hline
\end{array}
$$

13
$$
\begin{array}{r}
702_{sixteen} \\
- \ 289_{sixteen} \\
\hline
\end{array}
$$

2
$$
\begin{array}{r}
759_{twelve} \\
- \ 364_{twelve} \\
\hline
\end{array}
$$

6
$$
\begin{array}{r}
6D4_{fourteen} \\
- \ 27A_{fourteen} \\
\hline
\end{array}
$$

10
$$
\begin{array}{r}
A08\,3_{sixteen} \\
- \ 2FF_{sixteen} \\
\hline
\end{array}
$$

14
$$
\begin{array}{r}
50\,4_{fifteen} \\
- \ 1EE_{fifteen} \\
\hline
\end{array}
$$

3
$$
\begin{array}{r}
84A_{sixteen} \\
- \ 2B3_{sixteen} \\
\hline
\end{array}
$$

7
$$
\begin{array}{r}
946_{twenty} \\
- \ 62F_{twenty} \\
\hline
\end{array}
$$

11
$$
\begin{array}{r}
2B0A_{twelve} \\
- \ 436_{twelve} \\
\hline
\end{array}
$$

15
$$
\begin{array}{r}
600_{eleven} \\
- \ \ \ 4_{eleven} \\
\hline
\end{array}
$$

4
$$
\begin{array}{r}
92C_{sixteen} \\
- \ 5A7_{sixteen} \\
\hline
\end{array}
$$

8
$$
\begin{array}{r}
80\,3_{twenty} \\
- \ 2AE_{twenty} \\
\hline
\end{array}
$$

12
$$
\begin{array}{r}
61B_{twelve} \\
- \ 22A_{twelve} \\
\hline
\end{array}
$$

Part 11 Addition and subtraction of different bases

1	124 + 101	(base five)	16	203 − 26	(base seven)
2	243 + 124	(base six)	17	1001 − 110	(base two)
3	325 − 142	(base seven)	18	203 + 13 + 123	(base four)
4	840 − 217	(base nine)	19	144 + 23 + 314	(base five)
5	324 + 52	(base six)	20	1011 + 110 + 1011	(base two)
6	1101 − 111	(base two)	21	406 − 37	(base eight)
7	2031 − 113	(base four)	22	904 − 167	(base ten)
8	4305 − 1046	(base eight)	23	367 + 52 + 906	(base twelve)
9	236 + 179 + 25	(base ten)	24	32 + 203 + 13	(base four)
10	215 + 45 + 104	(base six)	25	1001 − 111	(base two)
11	1204 − 533	(base twelve)	26	431 − 98	(base sixteen)
12	10110 − 1101	(base two)	27	241 + 123 + 44	(base five)
13	415 − 66	(base eight)	28	101101 + 10111	(base two)
14	173 + 69 + 247	(base sixteen)	29	2000 − 111	(base three)
15	25 + 120 + 534	(base six)	30	437 − 188	(base eleven)

Systems of units

	Metric	**Non-metric**
Length	1 cm = 10 mm 1 m = 100 cm 1 km = 1000 m	1 foot = 12 inches 1 yard = 3 feet 1 chain = 22 yards 1 furlong = 10 chains 1 mile = 8 furlongs 1 fathom = 6 feet
Area	$1 \text{ cm}^2 = 100 \text{ mm}^2$ 1 are = 100 m^2 1 hectare = 100 ares	1 sq ft = 144 sq in 1 sq yd = 9 sq ft 1 acre = 4840 sq yd
Volume/Capacity	$1 \text{ cm}^3 = 1000 \text{ mm}^3$ 1 litre = 1000 cm^3	1 cu ft = 1728 cu in 1 cu yd = 27 cu ft 1 gallon = 8 pints
Mass	1 gram = 1000 mg 1 kg = 1000 grams 1 tonne = 1000 kg	1 lb = 16 oz 1 stone = 14 lb 1 cwt = 112 lb 1 ton = 20 cwt
Time	1 minute = 60 seconds 1 hour = 60 minutes 1 day = 24 hours 1 week = 7 days	

Questions

1 How long is a cricket pitch?

2 In what sport are furlongs still used?

3 What has a furlong to do with ploughing?

4 What does the letter 'c' stand for in cwt?

5 Which room in a house is most likely to have an instrument marked in stones as well as kilogrammes?

6 What connection is there between the pound (lb) and being born in September?

7 What are the origins of the words 'ounce' and 'inch'?

8 What do these prefixes mean?
   a milli-   b centi-   c kilo-   d hecto-

9 Which race of people first divided hours and minutes into 60 parts?

10 What connection is there between a metre and the North Pole and the Equator?

11 Approximately how many
   a cm in 1 inch   b inches in 1 metre
   c pints in 1 litre   d grams in 1 pound
   e litres in 1 gallon   f pounds in 1 kilogram
   g kilometres in 5 miles   h acres in 1 hectare?

12 Do you know of any other units not mentioned in the above table?

Systems of units

Part 1 Addition

1

ft	in
1	6
+	9

2

gall	pt
3	6
+ 2	5

3

yd	ft
1	2
+ 3	2

4

lb	oz
2	8
+ 4	12

5

min	sec
2	45
+ 6	25

6

gall	pt
2	5
1	6
+ 3	7

7

lb	oz
2	8
3	12
+ 1	14

8

yd	ft	in
2	1	8
+ 3	2	9

9

h	min	sec
1	42	30
+ 2	30	45

10

st	lb	oz
1	7	9
+ 2	9	8

11

days	h	min
3	18	40
+ 2	12	50

12

yd	ft	in
1	2	10
+ 2	2	9

13

st	lb	oz
1	8	7
+ 1	9	6

14

m	cm	mm
2	25	8
+ 1	63	7

15

yd	ft	in
4	1	10
+ 3	1	2

16

gall	pt
1	4
3	6
+ 2	7

17

min	sec
2	40
6	30
+ 12	55

18

yd	ft	in
1	2	8
2	1	9
+ 3	1	11

19

m	cm	mm
8	75	9
2	24	3
+ 1	18	9

20

days	h	min
1	16	50
2	20	30
+ 1	8	50

Part 2 Subtraction

1

ft	in
8	4
− 3	6

2

ft	in
5	6
− 2	10

3

gall	pt
4	3
− 1	7

4

gall	pt
6	5
− 3	6

5

lb	oz
6	4
− 2	10

6

lb	oz
9	9
− 5	12

7

min	sec
4	46
− 2	21

8

min	sec
10	25
− 6	45

9

yd	ft	in
4	2	6
− 2	1	8

Systems of units

10	yd	ft	in
	6	1	8
−	4	1	11

11	h	min	sec
	3	50	10
−	1	32	50

12	m	cm	mm
	4	50	5
−	1	38	8

13	st	lb	oz
	3	8	10
−	1	12	6

14	days	h	min
	4	11	30
−	2	15	55

15	weeks	days	h
	3	2	9
−	1	6	16

16	yd	ft	in
	6	2	8
−	3	1	10

17	kg	g
	8	250
−	2	800

18	st	lb	oz
	4	2	6
−	2	10	10

19	yd	ft	in
	5	0	8
−	2	1	9

20	weeks	days	h
	2	4	10
−	1	6	20

Part 3 Multiplication

1	ft	in
	2	5
×		3

2	ft	in
	1	5
×		8

3	gall	pt
	4	5
×		4

4	gall	pt
	3	6
×		7

5	h	min
	2	20
×		7

6	lb	oz
	2	9
×		4

7	lb	oz
	1	10
×		5

8	yd	ft
	4	2
×		5

9	gall	pt
	4	$6\frac{1}{2}$
×		6

10	cm	mm
	2	6
×		7

11	yd	ft	in
	1	2	8
×			6

12	days	h	min
	1	8	20
×			4

13	kg	g
	2	450
×		5

14	yd	ft	in
	1	1	6
×			8

15	m	cm	mm
	2	41	5
×			6

16	yd	ft	in
	4	1	$10\frac{1}{2}$
×			8

17	weeks	days	h
	2	5	15
×			4

18	st	lb	oz
	1	9	6
×			3

19	km	m	cm
	2	305	45
×			6

20	h	min	sec
	12	45	35
×			5

Systems of units

Part 4 Division

1 2)3 ft 2 in

2 5)12 ft 6 in

3 7)15 gall 6 pt

4 4)10 gall 4 pt

5 2)5 yd 1 ft

6 6)7 h 12 min

7 8)25 lb 8 oz

8 4)9 cm 6 mm

9 5)17 lb 8 oz

10 7)9 h 27 min

11 3)14 kg 400 g

12 8)11 yd 2 ft 4 in

13 5)12 yd 1 ft 11 in

14 3)7 days 5 h 30 min

15 5)6 st 8 lb 3 oz

16 8)9 h 14 min 40 sec

17 12)38 m 42 cm 4 mm

18 6)3 yd 1 ft 6 in

19 7)3 kg 640 g

20 4)10 days 6 h 12 min

Part 5 A mixture

1
lb	oz
2	9
+ 3	8

2
yd	ft	in
1	2	9
+ 2	1	4

3
weeks	days	h
2	3	12
+ 1	5	14

4
gall	pt
6	5
− 4	7

5
h	min
4	15
− 2	40

6
lb	oz
6	3
− 2	12

7
gall	pt
3	2
×	5

8
cm	mm
2	4
×	4

9
yd	ft
3	2
×	6

10
h	min
2	25
×	4

11 2)5 ft 8 in

12 3)7 lb 8 oz

13 5)8 cm 5 mm

14
yd	ft	in
2	1	9
+ 1	2	8

15
st	lb	oz
4	6	9
+ 8	9	12

16
gall	pt
8	2
− 5	7

17
h	min
8	40
− 1	28

18
yd	ft	in
2	1	8
×		3

Systems of units

19

	kg	g
	3	150
×		5

20 2)3 yd 2 ft 6 in

21 3)4 weeks 3 days 3 h

22

	st	lb	oz
	8	8	10
−	2	9	4

23

	km	m
	3	450
−	1	800

24

	h	min	sec
	1	25	30
×			3

25

	st	lb	oz
	2	3	6
×			4

26

	yd	ft	in
	1	2	8
	2	1	9
+	1	1	7

27

	cm	mm
	4	6
	12	8
+	11	7

28 8)9 min 12 sec

29 3)2 st 10 lb 4 oz

30

	yd	ft	in
	1	2	11
×			5

Part 6

All the calculations below have the correct answers.
Write the appropriate units of measurement for the head of each column.

1

+	1	5
	2	4
	4	1

2

+	4	2
	1	1
	6	0

3

+	2	8
	3	9
	6	1

4

+	1	9
		7
	2	4

5

+	6	4
	12	9
	19	3

6

+	2	30
	5	45
	8	15

7

×	2	6
		3
	7	2

8

×	2	2
		5
	13	1

9

×	12	4
		3
	37	2

10

×	3	5
		4
	13	4

11

×	3	5
		4
	14	4

12

×	1	3
		5
	7	1

13

−	4	2
	1	5
	2	5

14

−	5	1
	3	2
	1	2

15

−	4	1
	2	8
	1	9

16

−	14	8
	6	9
	7	9

17

−	3	5
	1	8
	1	9

18

−	4	10
	1	25
	2	45

19 2)9 2   (answer 4 5)

20 3)7 2   (answer 2 6)

21 5)6 2   (answer 1 1)

22 8)45 6   (answer 5 7)

23 4)10 2   (answer 2 4)

24 12)2 400   (answer 0 200)

Systems of units

Part 7 Problems

1. A gardener has three bags of peat with masses of 6 lb 8 oz, 2 lb 12 oz and 15 lb 4 oz. How much has he altogether?

2. A farmer has three buckets holding 2 gall 3 pt, 4 gall 5 pt and 1 gall 7 pt of milk. How much milk has he altogether?

3. Mr Jones had three holidays last year lasting 3 weeks 2 days, 2 weeks 4 days and 1 week 3 days. How long did he spend on holiday altogether?

4. A length of wire is 9 ft 4 in long. If 7 ft 10 in is cut off, how much is left?

5. A train takes 4 h 30 min to travel between two cities. A man goes by car and takes 6 h 10 min. How much longer does the journey take by car?

6. Before Mr Thomson went on a diet, he used to have a mass of 15 st 6 lb. He lost 3 st 10 lb. What is his mass now?

7. A metal can will hold 5 gall 4 pt of petrol. How much will six similar cans hold altogether?

8. It takes 2 h 25 min to fly from Manchester to Frankfurt. If a business man makes this trip six times, how long does he spend in flight?

9. A row of five houses all have gardens of the same width. If the width of one garden is 8 yd 2 ft, find how far it is across all five gardens.

10. A man spends 7 h 15 min at work dividing his time equally between three jobs. How long does he spend on each job?

11. Four cakes of the same size are made from a total of 9 lb 12 oz of flour. How much flour went into each cake?

12. A light fitting needs three pieces of electrical wire all the same length. One piece 8 yd 2 ft long is cut into three. How long will each piece be?

13. A factory worker's time-card shows that she works each day from Monday to Friday for these hours:
8 h 20 min, 7 h 50 min, 8 h 54 min, 6 h 12 min, 8 h 34 min.
Find the total time she spent at work that week?

14. A tin of peaches has a mass of 1 lb 3 oz. If 12 tins are placed into a box with a mass of 14 oz, what is
   a the mass of all the tins
   b the total mass of tins and box?

15. An automatic machine takes 2 min 35 sec to make a certain component. It takes 4 min 20 sec to warm up after being switched on. From being switched on, how long does it take to produce 8 components?

16. Five boxes are placed on a shelf side by side. The widths of the boxes are 2 ft 6 in, 1 ft 10 in, 3 ft 8 in, 1 ft 2 in, 2 ft 8 in. If the shelf is 4 yd 1 ft 6 in long, find the length of shelf which is empty?

17. Between 8 a.m. and 12 noon, a housewife washes three loads in her washing machine, lasting 55 min, 1 h 12 min, and 1 h 25 min. For how long between 8 a.m. and 12 noon is the washing machine *not* operating?

18. 6 gall 5 pt of milk has 1 gall 4 pt of water added and the total quantity is poured into 5 cartons of equal size. How much will each carton hold?

19. Mrs Netherton has two packets of flour with masses of 4 lb 8 oz and 5 lb 12 oz. If she used all this flour to make four loaves of bread of the same size, how much flour would be used in each loaf?

Systems of units

20 Eight similar books are stacked in a cardboard box 2 ft 9 in deep. If each book is $2\frac{1}{2}$ inches thick, what depth of the box is still empty?

21 A rectangle is 2 yd 1 ft 8 in long and 1 yd 2 ft 7 in wide. Find the perimeter of the rectangle.

22 Twelve lengths of telephone wire, each 85 metres long, are strung alongside a road. If the road is 1 km 450 m long, what length of it has *no* telephone wire along it?

23 Mrs Campbell can weave 1 ft 4 in of cloth each hour.
   a What length of material does she weave in a full 7-day week, working for 4 hours each day?
   b If this length is cut into eight equal pieces, how long is each piece?

24 A metal stake 1 ft 4 in long is driven into the trunk of a tree, so that 11 in of the stake still shows. If the tree trunk is 3 ft 2 in across, what width of the trunk has the stake *not* penetrated?

Binary numbers

Part 1 Paper tapes

Binary numbers can be fed into some computers on paper tape, where a hole stands for the number 1, and no hole (a blank) stands for a 0.

Copy and complete this table.

The five columns headed **Paper tape** show the holes and blanks to represent the five place values of the binary numbers.

Letter	Base ten number	Binary number 16	8	4	2	1	Paper tape				
A	1					1					○
B	2				1	0				○	
C	3				1	1				○	○
D	4			1	0	0		○			
E	5										
F	6										
G	7										
H	8										
I	9										
J	10										
K	11										
L	12										
M	13										
N	14										
O	15										
P	16										
Q	17										
R	18										
S	19										
T	20										
U	21										
V	22										
W	23										
X	24										
Y	25										
Z	26										
space	27										
fullstop	28										
comma	29										
?	30										
delete	31										

Use the table to help you decode the messages on the paper tapes on page 32.

Binary numbers

Tape A	Tape B	Tape C	Tape D

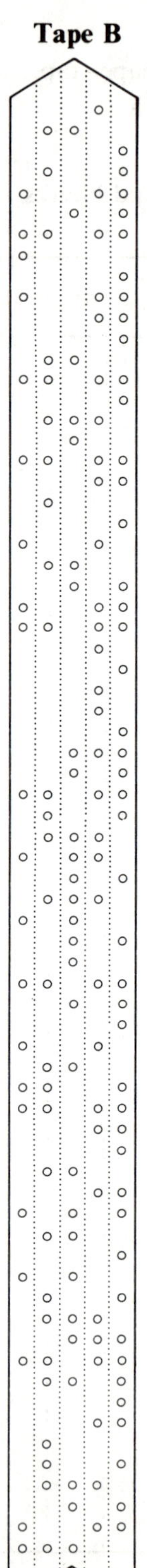
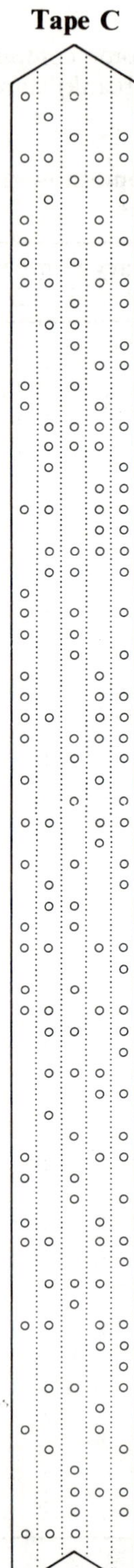
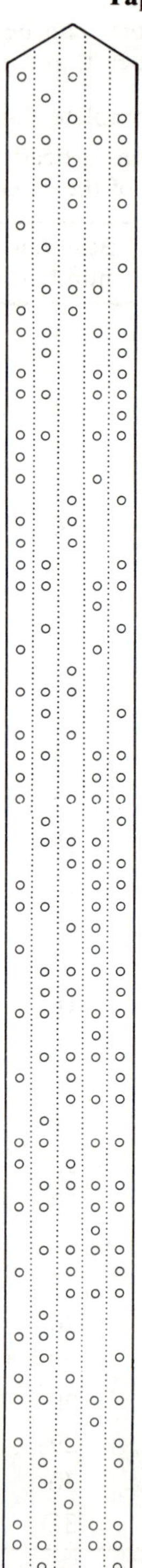
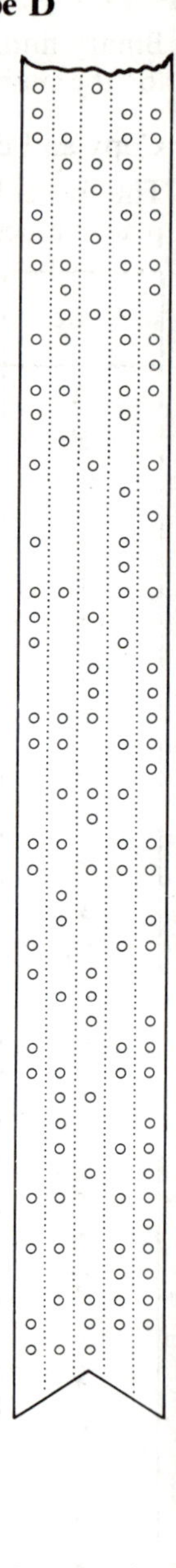

Now write your own messages on a paper tape and get a friend to decode them.

Binary numbers

Part 2 Switching circuits

Switches in an electrical circuit can be used to represent binary numbers.

    0 stands for a switch **open**.
    1 stands for a switch **closed**.

If current can **flow** through the circuit, then write a 1 in the flow column.
If current can**not flow** through the circuit, then write a 0 in the flow column.

Copy the table for each diagram and complete the last column.
(Note that the battery is omitted from the diagrams.)

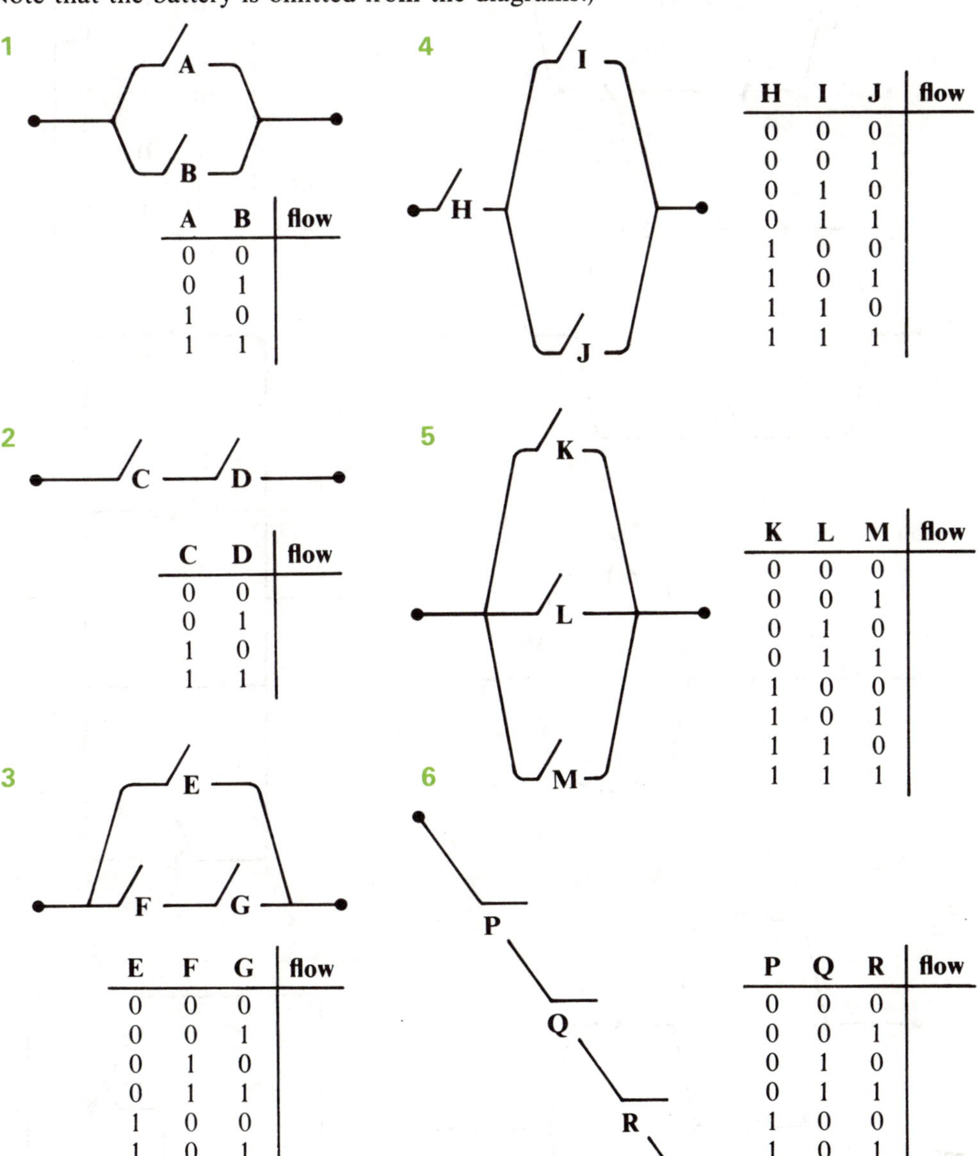

1

A	B	flow
0	0	
0	1	
1	0	
1	1	

2

C	D	flow
0	0	
0	1	
1	0	
1	1	

3

E	F	G	flow
0	0	0	
0	0	1	
0	1	0	
0	1	1	
1	0	0	
1	0	1	
1	1	0	
1	1	1	

4

H	I	J	flow
0	0	0	
0	0	1	
0	1	0	
0	1	1	
1	0	0	
1	0	1	
1	1	0	
1	1	1	

5

K	L	M	flow
0	0	0	
0	0	1	
0	1	0	
0	1	1	
1	0	0	
1	0	1	
1	1	0	
1	1	1	

6

P	Q	R	flow
0	0	0	
0	0	1	
0	1	0	
0	1	1	
1	0	0	
1	0	1	
1	1	0	
1	1	1	

Binary numbers

This table can be used with each of these circuits.

Copy the last column (the flow column) only for each diagram.

W	X	Y	Z	flow
0	0	0	0	
0	0	0	1	
0	0	1	0	
0	0	1	1	
0	1	0	0	
0	1	0	1	
0	1	1	0	
0	1	1	1	
1	0	0	0	
1	0	0	1	
1	0	1	0	
1	0	1	1	
1	1	0	0	
1	1	0	1	
1	1	1	0	
1	1	1	1	

7

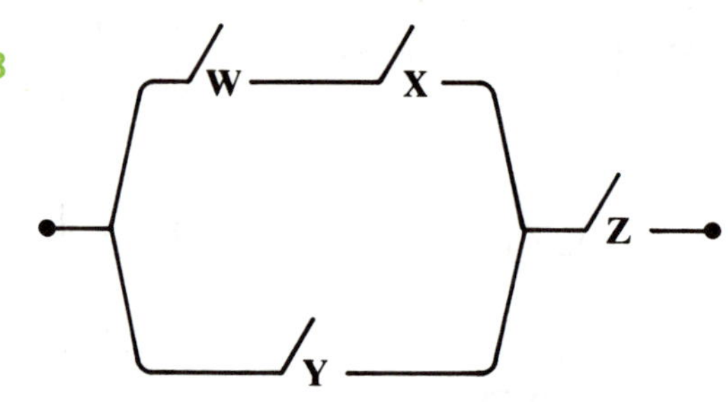

8

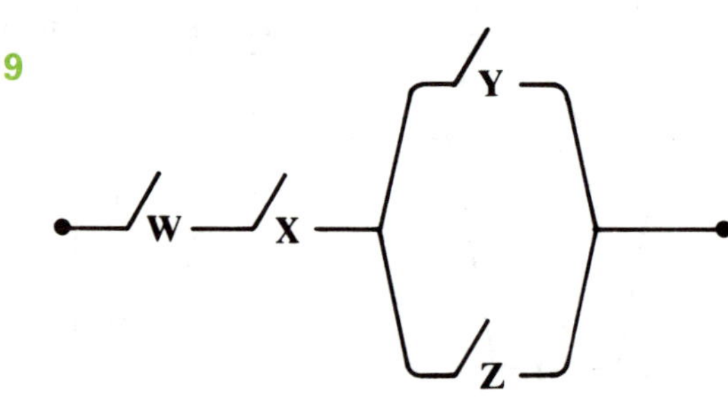

9

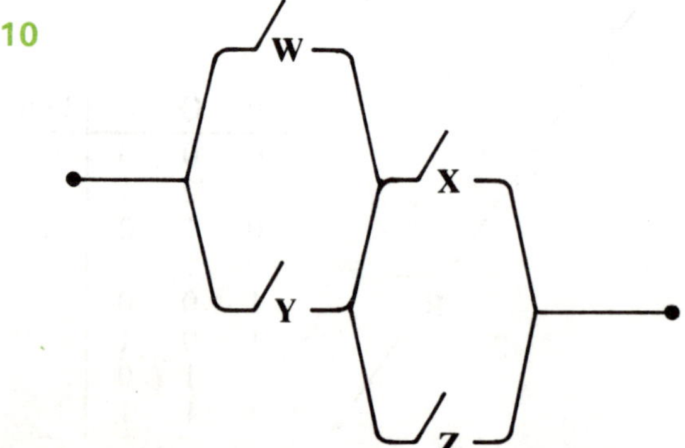

11

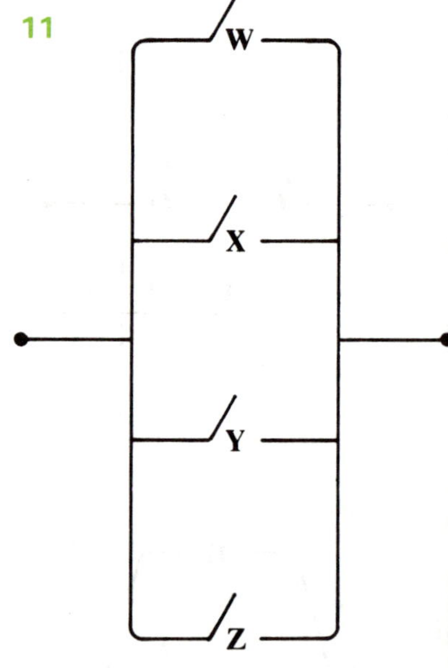

10

12

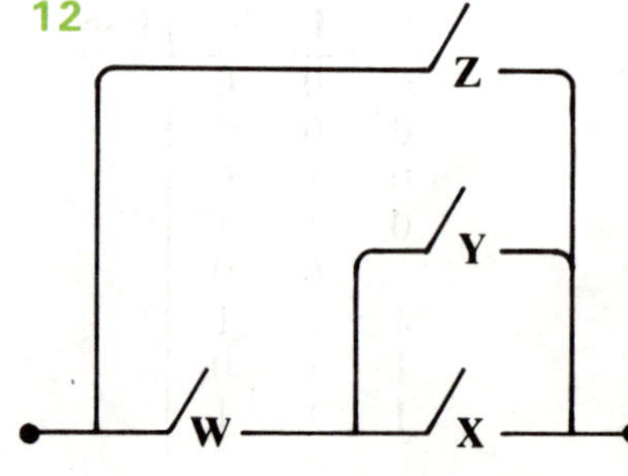

Binary numbers

Part 3 Railway marshalling yards

Each waggon of a train is given a binary number which informs the control box how to set the points. The waggon runs over the hump and under gravity into the correct siding.

Each 0 sets the points for **straight over**.
Each 1 sets the points for **turn aside**.

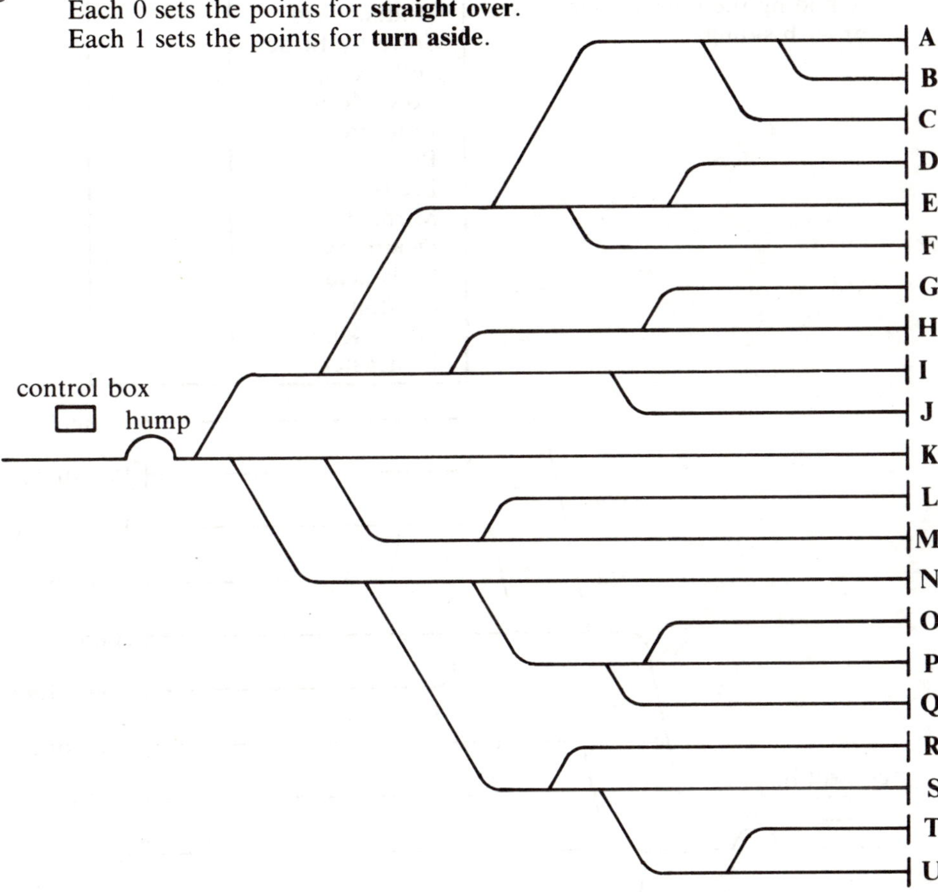

1 Copy this table and complete the last column with the letter of the correct siding taken from the above diagram.

City/Town	Binary code	Letter	City/Town	Binary code	Letter
Birmingham	000		Crewe	11000	
Grimsby	0010		Cardiff	01100	
Bristol	0111		Edinburgh	11100	
Hull	0011		Southampton	01011	
Holyhead	1000		Portsmouth	010100	
Doncaster	1010		Glasgow	11101	
Newcastle	1101		Preston	11001	
Shrewsbury	1001		Exeter	011011	
Nottingham	0100		London	010101	
York	1011		Penzance	011010	
Carlisle	1111				

Binary numbers

2 The sidings shown below are for waggons going to areas in Yorkshire.
Copy and complete this table by finding the binary code for each siding.

City/Town	Binary code
Sheffield	
Penistone	
Huddersfield	
Sowerby Bridge	
Halifax	
Low Moor	
Bradford	
Pudsey	
Leeds	
Mirfield	
Dewsbury	
Wakefield	
Crofton	
South Kirby	
Doncaster	

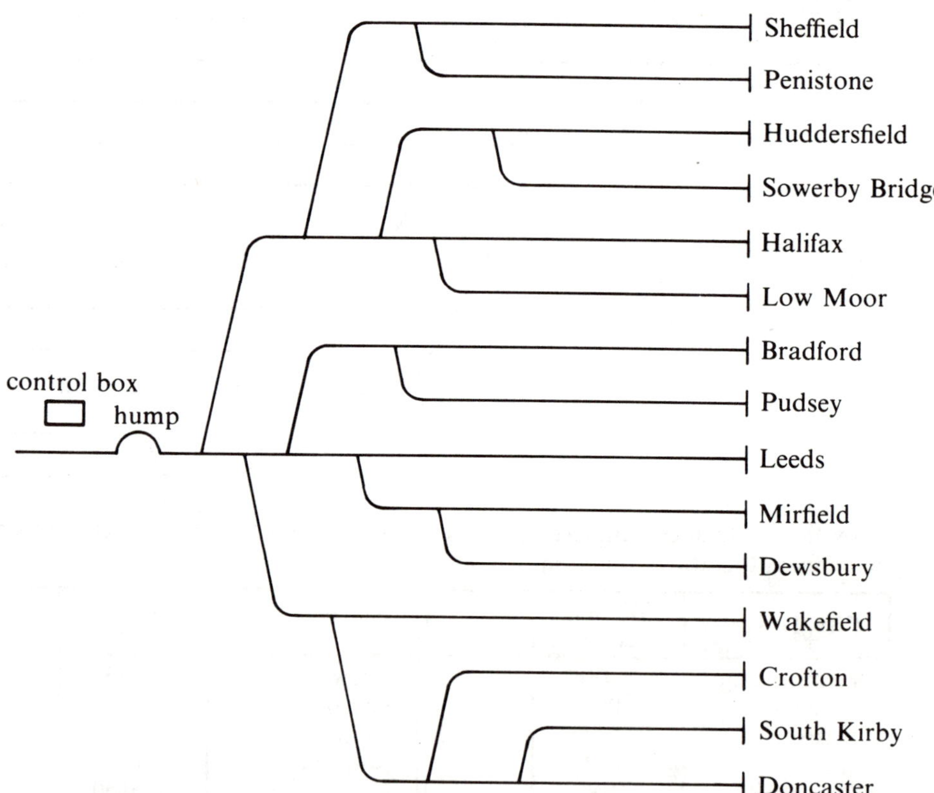

3 A small marshalling yard has only six sidings with the codes given here. Design the layout of the sidings for these codes and label each with its town's name.

Preston	00	Lytham	10
Leyland	010	Fleetwood	110
Southport	011	Blackpool	111

Binary numbers

Part 4

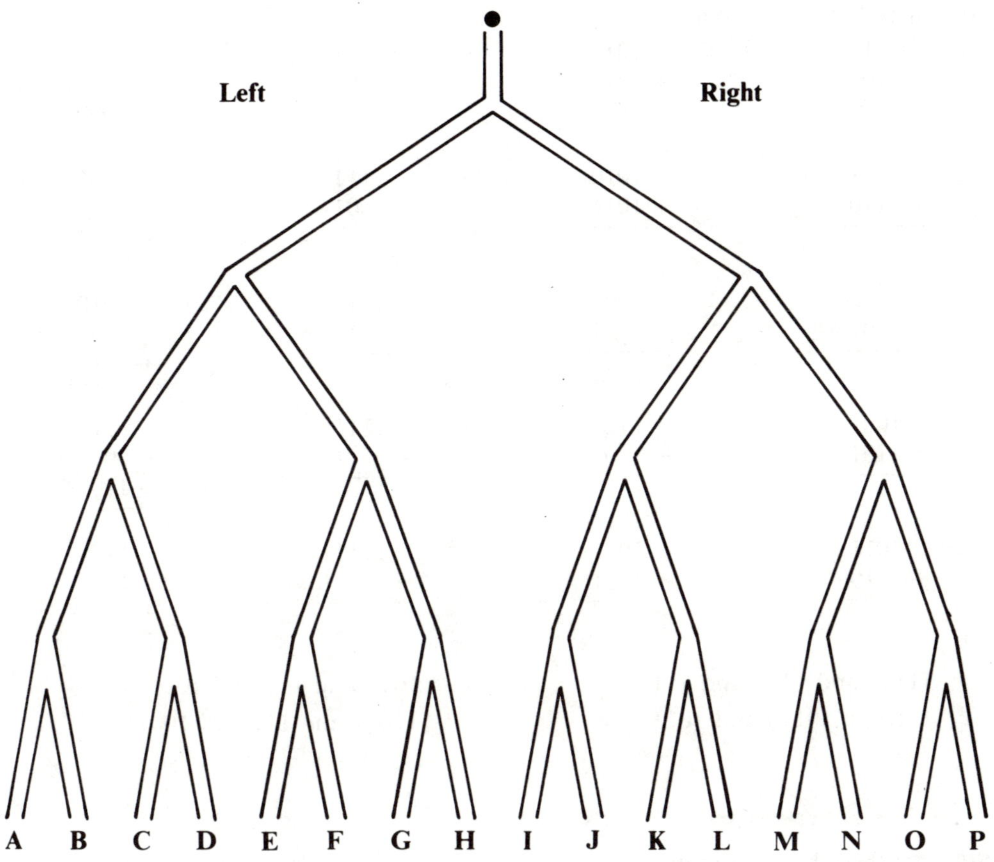

A small ball is placed at the entrance to this arrangement of tubes.

At each junction it can turn to the left or to the right as labelled in the diagram.

Find which binary numbers give the codes for the exits A to P, using

0 if the ball turns to the left and 1 if the ball turns to the right.

Copy the table and write in the binary numbers.

Convert each binary number into base ten.

What do you notice?

Exit	Binary number	Base ten number
A		
B		
C		
D		
E		
F		
G		
H		
I		
J		
K		
L		
M		
N		
O		
P		

Binary numbers

Part 5 Addition

Add these binary numbers.

1 1011 + 100	**6** 101 + 101	**11** 1101 + 1101	**16** 1010 1001 + 10001
2 1001 + 110	**7** 1010 + 1011	**12** 111 + 1001	**17** 1001 101 + 11
3 1010 + 101	**8** 1001 + 11	**13** 111 + 1	**18** 1010 1101 + 11
4 1001 + 101	**9** 111 + 10	**14** 1111 + 1	**19** 111 101 + 110
5 1011 + 10	**10** 1011 + 1011	**15** 11111 + 1	**20** 11 10 + 11

21   11101 and 1101 and 111

22   111 and 10110 and 1011

23   1011 and 111 and 1011

24   11011 and 111 and 10011

Part 6 Subtraction

Subtract these binary numbers.

1 1111 − 101	**6** 1010 − 101	**11** 10110 − 1001	**16** 10110 − 1101
2 1101 − 101	**7** 1101 − 111	**12** 10010 − 1001	**17** 100 − 1
3 110 − 1	**8** 1011 − 110	**13** 1110 − 111	**18** 100 − 11
4 101 − 11	**9** 1100 − 110	**14** 11011 − 1110	**19** 1000 − 110
5 1011 − 101	**10** 110110 − 11011	**15** 11101 − 1011	

20   Subtract 101 from 1100.

21   Subtract 110 from 1011.

22   Subtract 11 from 10101.

23   Subtract 1111 from 10000.

24   Subtract 11 from 10010.

Binary numbers

Part 7 Multiplication

Multiply these binary numbers.

1 $\begin{array}{r} 110 \\ \times\ 10 \\ \hline \end{array}$	**5** $\begin{array}{r} 110 \\ \times\ 11 \\ \hline \end{array}$	**9** $\begin{array}{r} 1011 \\ \times\ 100 \\ \hline \end{array}$	**13** $\begin{array}{r} 1001 \\ \times\ 111 \\ \hline \end{array}$

1 $\begin{array}{r} 110 \\ \times\ 10 \\ \hline \end{array}$ **5** $\begin{array}{r} 110 \\ \times\ 11 \\ \hline \end{array}$ **9** $\begin{array}{r} 1011 \\ \times\ 100 \\ \hline \end{array}$ **13** $\begin{array}{r} 1001 \\ \times\ 111 \\ \hline \end{array}$

2 $\begin{array}{r} 100 \\ \times\ 11 \\ \hline \end{array}$ **6** $\begin{array}{r} 111 \\ \times\ 11 \\ \hline \end{array}$ **10** $\begin{array}{r} 1111 \\ \times\ 100 \\ \hline \end{array}$ **14** $\begin{array}{r} 1101 \\ \times\ 111 \\ \hline \end{array}$

3 $\begin{array}{r} 101 \\ \times\ 11 \\ \hline \end{array}$ **7** $\begin{array}{r} 101 \\ \times\ 111 \\ \hline \end{array}$ **11** $\begin{array}{r} 1011 \\ \times\ 101 \\ \hline \end{array}$ **15** $\begin{array}{r} 11011 \\ \times\ 111 \\ \hline \end{array}$

4 $\begin{array}{r} 1010 \\ \times\ 11 \\ \hline \end{array}$ **8** $\begin{array}{r} 101 \\ \times\ 101 \\ \hline \end{array}$ **12** $\begin{array}{r} 1101 \\ \times\ 101 \\ \hline \end{array}$

16 Multiply 11101 by 100.

17 Multiply 101101 by 1000.

18 Multiply 10101 by 101.

19 Multiply 110011 by 1100.

20 Multiply 11011 by 1011.

21 Multiply 10111 by 10010.

22 Multiply 10110 by 1101.

23 Multiply 11101 by 111.

24 Multiply 11111 by 11111.

Part 8 Division

Change these binary numbers into base ten to divide but give the answers as binary numbers. Alternatively, all the working can be done in base two.

1 Divide 11011 by 11.

2 Divide 1110 by 10.

3 Divide 10100 by 101.

4 Divide 10100 by 100.

5 Divide 11110 by 110.

6 Divide 1111 by 101.

7 Divide 11011 by 1001.

8 Divide 11001 by 101.

9 Divide 11110 by 101.

10 Divide 110010 by 101.

11 Divide 110110 by 110.

12 Divide 111111 by 1001.

13 Divide 110110 by 1001.

14 Divide 1001000 by 1100.

15 Divide 1001011 by 11.

16 Divide 1100100 by 101.

17 Divide 1110101 by 1001.

18 Divide 1001011 by 11001.

19 Divide 1111110 by 111.

20 Divide 111100 by 1100.

21 Divide 1111110 by 1001.

22 Divide 11011011 by 1001001.

23 Divide 10011111 by 110101.

24 Divide 11111111 by 110011.

Binary numbers

Part 9 Equations

Solve these equations where all numbers are in base two. Give the answers in the same base.

1 $x - 101 = 111$

2 $x - 110 = 1011$

3 $x + 11 = 1010$

4 $x + 111 = 10110$

5 $11x = 10101$

6 $101x = 11001$

7 $\dfrac{x}{10} = 101$

8 $\dfrac{x}{11} = 111$

9 $101x - 11 = 10001$

10 $1100x + 101 = 1000001$

11 $11(x + 111) = 11110$

12 $101(x + 1001) = 110111$

13 $10(x - 110) = 1000$

14 $1010(x - 111) = 101000$

15 $11(10x - 111) = 1001$

16 $110(11x - 1010) = 1100$

Ratio and Proportion

Ratio
The unitary method
Tables and graphs for proportion

Ratio

Part 1

1 Write these ratios as simply as possible.

a	black dots to white dots	
b	ticks to crosses	
c	a's to z's	
d	squares to circles	
e	m's to n's	
f	ticks to crosses	
g	£ signs to $ signs	
h	6's to 9's	

2 For each diagram, write as simply as possible the ratio of shaded pieces to unshaded pieces.

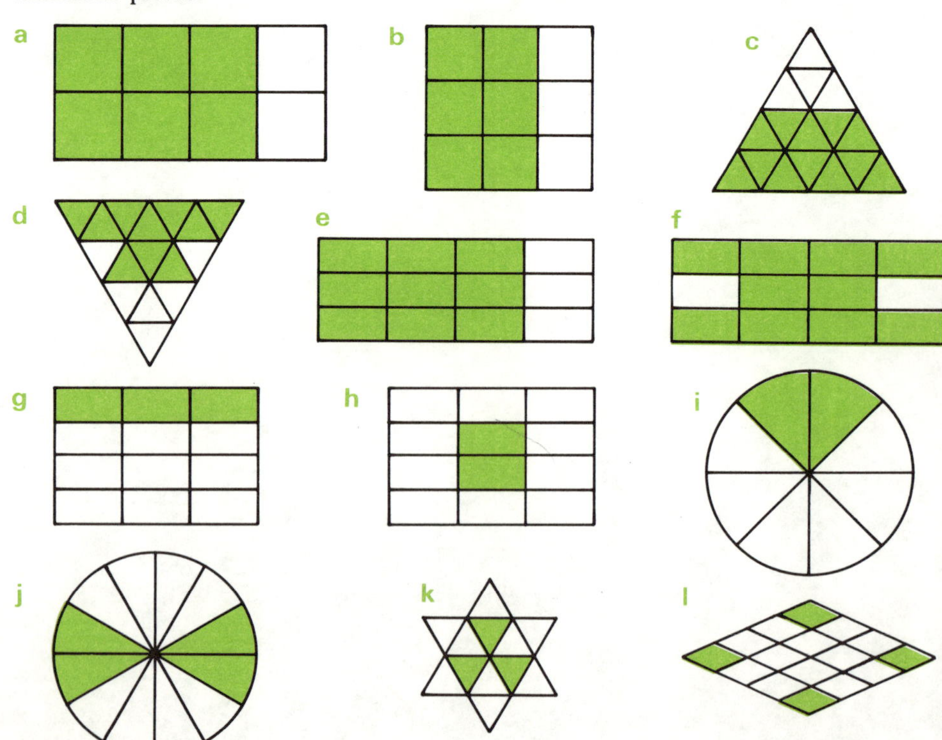

Ratio

3   Write these ratios in their simplest forms.

a	$2:8$	b	$4:6$	c	$7:21$	d	$10:15$
e	$9:12$	f	$6:18$	g	$3:12$	h	$8:4$
i	$15:5$	j	$36:6$	k	$35:10$	l	$18:12$
m	$\frac{1}{2}:3$	n	$\frac{1}{4}:5$	o	$\frac{1}{3}:2$	p	$\frac{3}{4}:5$
q	$\frac{2}{3}:1$	r	$3:\frac{2}{5}$	s	$\frac{1}{4}:\frac{1}{2}$	t	$\frac{2}{3}:\frac{5}{6}$
u	$\frac{3}{4}:\frac{1}{2}$	v	$\frac{5}{8}:\frac{1}{2}$	w	$\frac{1}{4}:\frac{3}{8}$	x	$\frac{7}{8}:\frac{3}{4}$

4   Express these ratios in the form $1:n$.

a	$2:10$	b	$3:15$	c	$7:14$	d	$12:48$
e	$2:5$	f	$2:15$	g	$4:10$	h	$3:10$
i	$5:11$	j	$5:12$	k	$4:31$	l	$4:2$
m	$6:3$	n	$9:3$	o	$8:6$	p	$6:4$

5   Find the value of each letter.

a	$1:4=3:n$	b	$1:6=5:m$	c	$1:9=p:27$
d	$2:5=q:10$	e	$3:4=r:20$	f	$6:1=42:s$
g	$5:2=t:8$	h	$12:4=3:u$	i	$16:8=v:1$
j	$36:9=w:1$	k	$64:24=x:3$	l	$7:y=35:40$

6   Of each of these pairs, which is the greater ratio?

a	$1:6$	$2:14$	b	$1:4$	$3:15$	c	$5:45$	$1:8$
d	$2:10$	$3:18$	e	$9:27$	$6:12$	f	$7:56$	$9:81$
g	$3:39$	$6:72$	h	$2:11$	$4:21$	i	$3:10$	$2:7$
j	$5:35$	$6:43$	k	$\frac{1}{2}:1$	$\frac{1}{3}:1$	l	$\frac{1}{4}:1$	$\frac{1}{3}:2$
m	$\frac{1}{8}:2$	$\frac{1}{5}:3$	n	$\frac{1}{9}:4$	$\frac{1}{7}:5$			

7   If $a:b:c=2:6:9$, find these ratios in their simplest terms.
   a $a:b$ b $b:c$

8   If $x:y:z=9:12:16$, find these ratios in their simplest terms.
   a $x:y$ b $y:z$

9   If $p:q:r=4:9:24$, find these ratios in their simplest terms.
   a $p:r$ b $q:r$

10   If $t:u:v=20:12:3$, find these ratios in their simplest terms.
   a $u:v$ b $t:u$

11   If $i:j:k=18:15:10$, find these ratios in their simplest terms.
   a $i:j$ b $j:k$ c $i:k$

12   If $l:m:n=20:15:12$, find these ratios in their simplest terms.
   a $l:m$ b $m:n$ c $l:n$

13   An alloy contains 15 kg of iron and 3 kg of copper. Find the ratio of iron to copper.

14   A man spends £8 on petrol and £2 on motor oil. Find the ratio of money spent on petrol to money spent on oil.

15   To make concrete, mix 8 buckets of sand with 2 buckets of cement. Write the ratio of sand to cement.

Ratio

16 A cake contains 480 grams of fat and 720 grams of flour. Find the ratio of fat to flour.

17 Martin James spent 45 minutes on his maths homework and 36 minutes on his geography. What is the ratio of the time spent on maths to geography?

18 My front garden is 12 metres long and my back garden is 15 metres long. What is the ratio of the length of the front garden to the back garden?

19 Two cogs interlock; one has 80 teeth and the other has 240 teeth. Find the ratio of their teeth (this is called the **GEAR RATIO**).

20 Find the gear ratio of two cogs. One has 120 teeth, the other 360 teeth.

21 What is the gear ratio of two cogs, one of which has 64 teeth and the other 192 teeth?

22 A triangle has sides of length 12 cm, 24 cm and 30 cm. Write the ratio of the three sides in its simplest form.

23 Another triangle has sides of lengths 15 cm, 18 cm and 24 cm. What is the ratio of the three sides in its simplest terms?

24 Three brothers go on holiday with £12, £16 and £20 pocket money. Find the ratio of these amounts.

25 Three sisters are given £1·20, £1·50 and £2·40 to spend. Find the ratio of these amounts as simply as possible.

26 a If $a : b = 3 : 4$, find the value of b when $a = 6$.

b If $c : d = 2 : 5$, find the value of d when $c = 8$.

c If $e : f = 6 : 7$, find the value of f when $e = 30$.

d If $g : h = 5 : 9$, find the value of h when $g = 25$.

e If $i : j : k = 2 : 3 : 4$, find the values of j and k when $i = 10$.

f If $l : m : n = 6 : 7 : 10$, find the values of m and n when $l = 18$.

g If $p : q : r = 1 : 2\frac{1}{2} : 3$, find the values of q and r when $p = 6$.

27 In a class at school, the boy to girl ratio is $1 : 2$. If there are 9 boys in the class, how many girls are there?

28 In another class in the same school, the boy to girl ratio is $2 : 3$. If there are 10 boys in the class, how many girls are there?

29 The ratio of homes without television to those with television is $1 : 20$. If on an estate, 3 homes have no television, how many will have television?

30 The ratio of private cars to company cars on British roads is $2 : 3$. If a car park has 40 private cars in it, how many company cars are likely to be parked there?

31 The ratio of pet cats to pet dogs is $5 : 6$. If there are 25 cats in a village, how many dogs should there be?

32 The teacher to pupil ratio in a primary school is $1 : 25$. If the school has 8 teachers, how many pupils has it?

33 The teacher to pupil ratio in a secondary school is $1 : 18$. If the school has 36 teachers, how many pupils has it got?

34 A machine makes plastic trays and the ratio of faulty ones to good ones is $1 : 30$. If it makes 4 faulty ones, how many good ones has it made?

35 Another machine makes plastic forks and its ratio of rejects to good ones is $3 : 16$. If 36 forks are rejected, how many are acceptable?

Ratio

36 The ratio of my gas bill to my electricity bill was 13 : 5. If my gas bill was £78, how much was my electricity bill?

37 The ratio of the speeds of two intermeshing cogwheels is 3 : 16. Find the speed of the faster wheel, if the slower one rotates at a speed of 24 revolutions per minute (rpm).

38 This pie chart shows how a family spends its income.

What is the ratio of its expenditure on
a travel to food
b house to food
c travel to house?

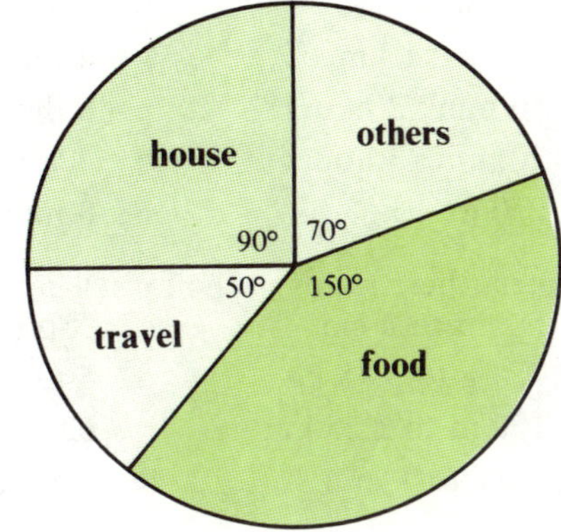

39 This pie chart shows where a class of pupils spent its summer holidays.

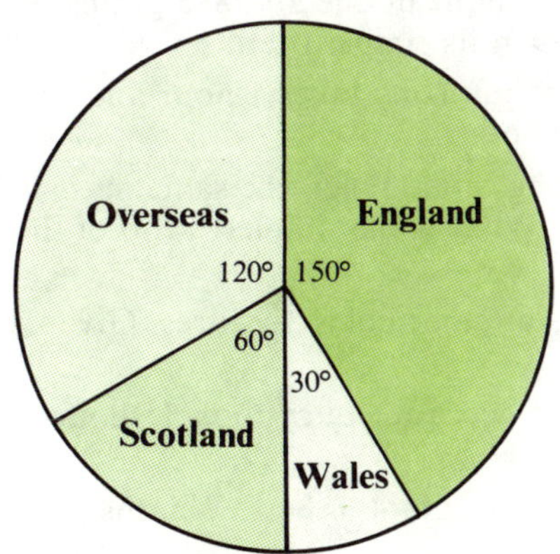

What is the ratio of pupils in
a Wales to Scotland
b Wales to England
c Scotland to Overseas
d Scotland to England
e Overseas to England?

If 12 pupils went on holiday in Wales, how many of them went on holiday
f in Scotland
g Overseas
h in England?

40 The pupils of a school have to choose one of the subjects shown in this pie chart, which gives the results of their choices.

What is the ratio of those choosing
a typing to metalwork
b metalwork to geography
c metalwork to history
d history to geography
e child care to geography?

If there are 240 pupils in the year, how many of them choose
f history
g typing
h geography?

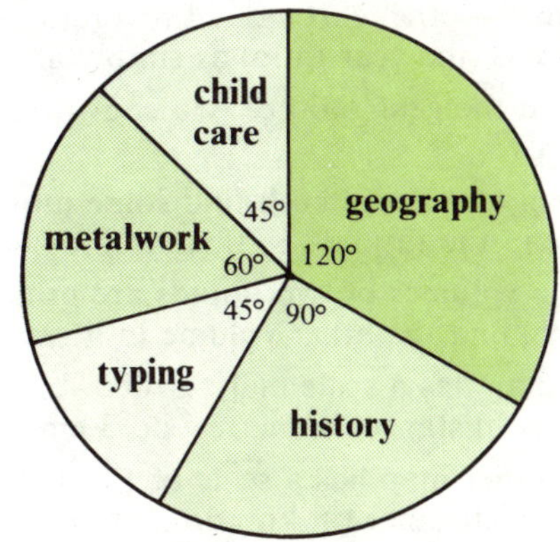

Ratio

Part 2

1   Write these ratios in their simplest forms.
   a   1 mm : 1 cm   b   1 day : 1 week   c   1p : £1
   d   2 mm : 1 cm   e   6 days : 2 weeks   f   2p : £1
   g   3 min : 1 hour   h   500 cm³ : 1 litre   i   100 g : 1 kg
   j   250 m : 1 km   k   4 mm : 2 cm   l   50p : £3
   m   30 min : 2 hours   n   6 days : 3 weeks   o   100 cm³ : 2 litres
   p   20 min : 3 hours   q   40p : £1·60   r   50p : £1·25
   s   250 g : ½ kg   t   8 mm : 3.6 cm   u   64p : £2·40

2   Express these ratios in the form 1 : n.
   a   5 mm to 4 cm   b   35p to £1·75   c   6 hours to 2 days
   d   750 g to 3.75 kg   e   5 min to 1½ hours   f   600 g to 1.8 kg
   g   150 m to 1.25 km   h   625 m to 2.5 km   i   75p to £6
   j   ½ sec to 1 min   k   ½ cm to 1 m   l   25 mm to 1 m

3   Peter gets £1·50 pocket money each week. Paul only gets 90p. Give the ratio of their pocket money in its simplest terms.

4   Mrs Cook bakes two cakes using 1.25 kg of flour in one and 875 grams of flour in the other. Give the ratio of these masses in its simplest terms.

5   Gary runs the cross-country in 50 minutes, but Tony takes 1 hour and 15 minutes. Give the simplest ratio of their times.

6   A spring extends 8 mm when a mass is hung on it. When the same mass is hung on another spring the extension is 2.8 cm. What is the simplest ratio of these extensions?

7   One bottle of wine holds 750 cm³ whereas another holds 1.2 litres. Give the simplest ratio of their capacities.

8   Jane took 45 min to do her homework, but her sister Lucy took 1¼ hours. What is the ratio of their times in its simplest form?

9   A girl has £1·08 in her purse and her friend has £1·44 in hers. What is the simplest ratio of these amounts?

10   Two cars are filled up with petrol. The cost for the first is £3·75 and for the other £8·75. Find the ratio of their costs.

11   Two cogs intermesh with one turning through 480 revolutions per minute (rpm) and the other through 24 revolutions per second (rps). Find the ratio of their speeds (the gear ratio) as simply as possible.

12   Find the gear ratio of two cogwheels which turn with speeds of 640 rpm and 8 rps.

13   Alan and Barry both find some money in the school playground. The amounts are in the ratio 3 : 8. If Alan finds 60p, how much does Barry find?

14   The volumes of two liquids are in the ratio 2 : 15. If the lesser volume is 160 cm³, find the other volume in litres.

15   A map has a scale of 1 : 20 000.
   What distance (in metres) do 3 cm on the map stand for?

16   Another map has a scale of 1 : 5 000 000.
   What distance (in km) do 3 cm on the map stand for?

Ratio

Part 3

1 Divide

a	£6·50	in the ratio 1 : 4
b	£8·56	in the ratio 1 : 3
c	£12·24	in the ratio 1 : 7
d	23.4 litres	in the ratio 1 : 5
e	47.1 metres	in the ratio 1 : 2
f	162.4 kg	in the ratio 1 : 6
g	£12·40	in the ratio 2 : 3
h	£14·91	in the ratio 2 : 5
i	£30·24	in the ratio 3 : 4
j	30.78 metres	in the ratio 4 : 5
k	47.16 metres	in the ratio 2 : 7
l	38.0 kg	in the ratio 3 : 5.

2 Divide

a	£31·56	in the ratio 1 : 2 : 3
b	£42·24	in the ratio 1 : 3 : 4
c	67.0 kg	in the ratio 1 : 2 : 7
d	31.05 kg	in the ratio 1 : 4 : 4
e	46.17 metres	in the ratio 1 : 2 : 6
f	18.4 litres	in the ratio 2 : 3 : 5
g	34.65 metres	in the ratio 2 : 4 : 5
h	669.6 kg	in the ratio 3 : 4 : 5.

3 The ratio of boys to girls in a class of pupils is 3 : 4. If there are 28 pupils in the group, find how many are
  a boys   b girls.

4 The ratio of boys to girls in a youth club is 4 : 5. If there are 63 members of the club, find how many are
  a boys   b girls.

5 A garage sells 468 cars (both estates and saloons). If the ratio of estates to saloons is 2 : 7, find how many
  a estates it sells   b saloons it sells.

6 Farmer Giles keeps cows and sheep in the ratio of 2 : 5 respectively. If he has 945 of these animals, how many cows and how many sheep has he?

7 John McDonald travels 275 miles to get to Stranraer. The ratio of distance on motorway to distance on ordinary roads is 3 : 8.
Find the distances he travels on
  a motorway   b ordinary roads.

8 The ratio of wet days to dry days in an English winter is 3 : 4. If 119 days are classed as 'winter', how many will you expect to be wet and how many dry?

9 A firm posts 1116 letters during a week. If the ratio of overseas mail to British mail is 2 : 7, find how many letters are sent
  a overseas   b within Britain.

Ratio

10 There are 520 houses on an estate and the ratio of those with double-glazing to those without it is 1 : 7. Find how many houses
   a have double-glazing   **b** do not have double-glazing.

11 For every 3 kg of butter a shop sells 7 kg of margarine.
   a Write the ratio of butter to margarine.
   b If a total of 150 kg of these products is sold, what mass is
      (i) butter   (ii) margarine?

12 For every 2 sixth-formers who have a Saturday job there are 5 that do not have one.
   a Write the ratio of those with jobs to those without.
   b If there are 126 sixth-formers, how many have Saturday jobs and how many do not?

13 For every 9 teenagers who like pop music there are 2 that do not.
   a Write the ratio of those who like pop music to those who do not.
   b In a youth club of 187 members, how many do not like pop music?

14 For every 5 people who go on holiday in August there are 3 people who do not.
   a Write the ratio of those who go on holiday in August to those who do not.
   b In a factory of 2760 people, how many people take their holidays in August?

15 The ratio of Mrs Park's gas bill to her electricity bill is 3 : 5. If she spent £97·20 altogether, how much did she spend
   a on gas   **b** on electricity?

16 Before a fight, two boxers are weighed and their masses are in the ratio of 4 : 5. If their combined mass is 174.6 kg, find the mass of each boxer.

17 A piece of cable is 128.1 metre long. It is cut into two sections with lengths in the ratio 3 : 4. Find the length of each section.

18 A farmer with 575.3 hectares of land divides it between arable and pasture in the ratios of 2 : 9. How many hectares are
   a arable   **b** pasture?

19 A lorry is loaded up with fruit and vegetables for market. The mass of fruit to vegetables is in the ratio of 7 : 8. If the lorry's load is 18.6 tonnes, find the mass of fruit and the mass of vegetables it is carrying.

20 If the ratio of the time a teacher spends teaching to the time he spends marking is 8 : 3, find how much time a teacher spends marking in a working week of 29.7 hours.

21 Divide £56·96   in the ratio 1 : 2 : 5.
22 Divide £170·52   in the ratio 1 : 4 : 7.
23 Divide 336 gallons   in the ratio 1 : 2 : 4.
24 Divide 77.13 metres   in the ratio 1 : 4 : 4.
25 Divide 26.28 metres   in the ratio 2 : 3 : 4.
26 Divide £51·92   in the ratio 2 : 4 : 5.
27 Divide 39.45 metres   in the ratio 4 : 5 : 6.
28 Divide 42.24 kilograms in the ratio 1 : 2 : 4 : 5.
29 Divide 60.72 kilograms in the ratio 1 : 3 : 3 : 4.
30 Divide £205·68   in the ratio 2 : 2 : 3 : 5.

Ratio

31 The profit of a business is £645 and this is divided between three men in the ratio 3 : 4 : 8. How much do they each receive?

32 In an exam, 165 marks are shared between three questions in the ratio 5 : 4 : 6. Find the marks for each question.

33 On the death of their grandfather, three sons inherit £6480 in the ratio 2 : 3 : 4 with the eldest son receiving most. How much does the youngest son inherit?

34 During the winter, three households use 224 bags of coal between them in the ratio 4 : 5 : 7. How much does each use?

35 An alloy consists of copper and nickel in the ratio 5 : 11. What mass of copper would you expect in 360 kg of alloy?

36 Another alloy consists of iron, copper, nickel and phosphorus in the ratio 7 : 4 : 3 : 1. How much
   a phosphorous   b nickel would you have in 219 kg of alloy?

37 Albert, Bertram and Claude provide £2500, £5000, £7500 respectively to set up a small business. Any profit they make is to be divided in the ratio of the money they each provided. If the first month's profits are £3240, how much does each receive?

38 The profit of a firm is divided in such a way that Alfred gets £4 for every £5 that Boris gets, and Cuthbert gets £7 for every £5 that Boris gets.
   a Write the ratio of their profits: Alfred to Boris to Cuthbert.
   b If the profit amounts to £8784, find how much they each receive.

39 Divide £42 into two parts so that one part is two-fifths of the other part.

40 Divide £65 into two parts so that one part is two-thirds of the other part.

The unitary method

Direct proportion

Part 1

1 In 6 hours a man earns £24. How much does he earn in 5 hours?

2 3 kg of sugar cost £1·20. How much do 4 kg cost?

3 A man walks 2 miles in 46 minutes. How long does he take to walk 3 miles?

4 4 packets of tea cost 84 pence. How much do 3 packets cost?

5 7 pencils cost 63 pence. How much do 12 pencils cost?

6 In 5 days my watch loses 20 minutes. How much does it lose in one week?

7 In 4 seconds a train travels 52 metres. How far does it travel in 3 seconds?

8 7 exercise books have a total mass of 224 grams. What is the mass of 6 of these books?

9 A pop group plays 8 songs in 36 minutes. How long would you expect 5 songs to take?

10 6 bags of flour have a total mass of 27 kg. What is the mass of 5 similar bags?

11 It takes 6 minutes to hard-boil 3 eggs in one pan. How long will it take to hard-boil 2 eggs in the same pan?

12 8 tubes of toothpaste have a total mass of 976 grams. What is the mass of 7 tubes?

13 9 washing-up liquid containers hold 2808 cm^3. What volume will 5 of these containers hold?

14 My car uses 8 litres of petrol over 104 km. How far will it travel on 11 litres?

15 If I buy 12 large envelopes for 180 pence, how much would I have to pay for 15 of them?

16 On school camp, 15 boys eat 345 slices of bread. If 18 boys had been camping, how many slices would you have expected them to eat?

17 4 buckets standing side-by-side are filled with rainwater in 32 minutes. If only 3 buckets had been there, how long would it have taken for them to fill?

18 In 12 minutes at a constant speed an aeroplane climbs to 27 000 metres. How high was it after 9 minutes?

19 In 24 hours the earth turns through 360 degrees. What angle does it turn through in 15 hours?

20 In 14 hours of steady driving a car travels 630 miles. How far did it go in 17 hours?

Part 2

1 If 6 peaches cost 84 pence, find the cost of 5 peaches.

2 A distance of 8 km is represented by 136 cm on a map. How many centimetres will represent 15 km?

3 12 m^2 of sheet metal have a mass of 276 kg. Find the mass of 14 m^2 of the same sheet.

4 In 20 minutes an aeroplane travels 260 miles. What distance will it travel in 32 minutes at the same speed?

5 15 cm^2 of a map represent a forest of 375 hectares. How many hectares will an area of 17 cm^2 on the map represent?

The unitary method

6 With 480 people in the audience, a violinist takes 16 minutes for his performance. How long would he have taken with only 320 people in the audience?

7 Over 21 weeks a boy saves £48·30 at a constant rate. How much did he save in the first 14 weeks?

8 The cost of hiring a car for 12 days is £61·92. How much would it have cost for only 7 days?

9 A lorry uses 11 litres of diesel to travel 137.5 km. How far will it travel on 16 litres?

10 A carpet of area 15 m² costs £64·80. How much would 23 m² of similar carpet cost?

11 £12 can be exchanged for 103.80 French francs. How many francs would you get for £34?

12 A force of 16 newtons will extend a spring by 52 cm. What distance will a force of 19 newtons extend it?

13 If it takes 22 minutes to empty a tank holding 272.8 m³ of liquid, how much liquid will have run out after 14 minutes?

14 With 144 passengers on board a train takes 16 minutes to travel between two stations. How long will it take with 132 passengers?

15 I exchange £45 for 182.7 DM before taking a holiday in Germany. If my wife changes £35, how many Deutschmark will she get?

16 A mast 22 metres high casts a shadow 52.8 metres long. How long would the shadow be from a flagpole 9.5 metres high at the same time?

17 It takes Mrs Smith's washing machine 18 minutes to wash 12 kg of clothes. How long will it take to wash 11 kg of clothes?

18 34 cm³ of a certain metal have a mass of 280.5 grams. What is the mass of 20.4 cm³ of this metal?

19 A 46-kilometre journey by train costs £5·75. How much will a 128-kilometre journey cost at the same rate?

20 A bicycle has a faulty cyclometer which records 24 miles for a trip of 39.6 miles. What distance is covered on another journey where the cyclometer records 32 miles?

Part 3

1 A workman earns £16·50 in 5 hours. How much does he earn in 4 hours?

2 A hiker walks 59.5 km in 7 hours. How far does he walk in 4 hours?

3 I receive 91.8 Swiss francs for £12 at the bank. How many S francs will I get for £14?

4 27.5 loaves of bread were eaten by 11 boys on a camping weekend. How many loaves would have been needed, if 17 boys had gone camping?

5 It took Mr and Mrs Gregson 3 hours to travel the 120 miles to Manchester Airport. If they had taken their 2 children with them, how long would it have taken?

6 A 495-page book is 15 mm thick. How many pages will there be in a book 23 mm thick, if the same type of paper is used?

The unitary method

7. A blue area of 51 cm² on a map represents a lake of 15 km². If a nearby lake has an area of 8.5 km², how many square centimetres of map will it take up?

8. I pay £91·35 for 21 bags of sand. How much will 17 bags cost?

9. A farmer ploughed 32.5 hectares of his land in 13 hours at a steady rate. How much had he ploughed after 7.5 hours?

10. Water from a tap takes 15 minutes to fill a 624-litre tank. What is the volume of a smaller tank, if water from the same tap took 11.5 minutes to fill it?

11. It takes Mrs McDonald 8 seconds to dial the 8-digit telephone number of her sister who lives 76 miles away in Manchester. How long will it take her to dial her brother's 8-digit number, if he lives 92 miles away in Liverpool?

12. My bicycle wheel goes round 32 times when I cycle 168 metres. How far do I go, if the wheel turns 132 times?

13. The front lawn of area 34 m² cost me £25 to turf. What is the area of my back lawn, if it cost £42·50 to turf?

14. A dripping tap fills a tin 36 mm deep in water in 16 minutes. How deep was the water after 14 minutes?

15. 980 grams of breakfast cereal will serve 35 people in a hotel. What mass of cereal is needed for 56 people?

16. A sewing machine puts 18 stitches into 24 mm of material. How many stitches will be sewn into a length of 148 mm?

17. A hot-air balloon rose 640 metres steadily in 16 minutes. How far did it rise in 9 minutes?

18. A funicular railway climbs 112 metres in a time of 35 seconds. How long does it take to climb 64 metres?

19. £127·50 was left in a bank for 5 years; during this time it gained £61·50 simple interest. How much interest had it earned in 3 years?

20. My car used 0.27 gallons of petrol over a test run of 12 miles. How much fuel will it need for a 204-mile trip?

Inverse proportion

Part 4

1. 6 boys can paint a fence in 4 days. How long will it take 8 boys?

2. 4 men lay a pipeline in 5 days. How long would 10 men take?

3. Travelling at 6 miles per hour a man takes 4 hours for a cycle trip. How long would he take at a speed of 8 mph?

4. Water from 4 taps will fill a water tank in 12 minutes. If one tap is closed how long would water from 3 taps take?

5. Two men can paint a room in 120 minutes. How long would 3 men take?

6. A quantity of oats will feed 6 horses for 10 days. For how long would the same quantity feed 4 horses?

7. An express train travelling at 80 km/h takes 3 hours for a journey. How long will a slower train take, if its speed is 60 km/h?

8. Mr Jones and his friend take 12 months to build a house. If there had been 6 men working on it, how long would it have taken?

The unitary method

9 5 pumps can empty a tank of water in 6 hours. If one broke down, how long would 4 pumps take?

10 A shelf is filled with 18 books, each of them 2 cm thick. If these books are replaced by 12 others of equal thickness, how thick will they each be to fill the shelf?

11 The wind and sun take 40 minutes to dry 6 towels on a line. How long would 4 towels have taken to dry?

12 Peaches are packed into trays. With 6 peaches per tray, 15 trays are needed. How many trays would be required with 9 peaches per tray?

13 A car with 3 passengers takes 4 hours to complete a certain journey. How long would the journey have taken with only 2 passengers?

14 You buy enough material to make 8 curtains 3 metres wide. How wide would they have been, if you had made 6 curtains?

15 I have enough bran to feed a litter of six puppies for 14 days. If there had been seven puppies, how long would the bran have lasted?

16 Bags of sugar are packed in 7 boxes each holding 12 bags. If only 6 boxes were used, how many bags would there be in each box?

17 A farmer needs to use 2 combines for 9 days to harvest all his corn. How long would it take, if he had 3 combines?

18 If I spend £1·50 a day, my holiday can last for 14 days. How long would it last, if I spent only £1·05 each day?

19 A sum of money is to be spent on tennis rackets. I can buy 12 rackets, if they cost £10·53 each. If I decide to buy 13 different rackets for the same total amount, how much would each cost?

20 A charity raises money to buy chairs for an old-persons' home. There is sufficient money to buy 9 ordinary chairs at £75·60 each. Alternatively, how many armchairs could be bought at £97·20 each?

Part 5

1 At 30 km/h, a train takes 6 hours to cover a certain distance. How long would it take at 36 km/h?

2 A load of hay will feed 12 horses for 15 days. For how long would it feed 10 horses?

3 It takes 9 men 40 days to mend a road. How long would the job take 15 men?

4 A shop assistant packs 18 tins into each box and uses 20 boxes. How many boxes will she need, if only 12 tins are put in each box?

5 9 men build a wall in 20 days. How long would 12 men take?

6 From the top of a hill two men can see 24 miles. How far would three men be able to see?

7 I buy some stamps and receive 12 rows with 18 stamps in each row. How many rows would there be, if there were only 8 stamps in each row?

8 A secretary can type 15 pages of 30 lines each in a morning. If she had typed 9 pages, how many lines on a page would there have been?

9 Travelling on the motorway, I took 45 minutes at a steady speed of 70 mph to complete a journey. At what speed would I have been travelling, if the same journey took me 50 minutes?

The unitary method

10 A florist puts 9 flowers into each bunch and makes up 60 bunches. How many bunches would there have been with 12 flowers in each?

11 Water from six taps fills an empty tank in 4 hours 15 minutes. How long would it have taken, if one tap had been turned off?

12 A flower-bed has space for 160 plants placed 9 cm apart. How many plants would it take, if they were 15 cm apart?

13 A truck is strong enough to carry 3240 bricks, each of mass 1.25 kg. What would be the mass of a different size of brick, if the truck could carry 2700 of them.

14 I can pay off my debts at £1·50 each week for 25 weeks. How long will it take, if I only pay £1·25 per week?

15 My car is fully loaded when I put 6 suitcases, each with a mass of 35.5 kg, on the roof rack. What would be the mass of each case, if 5 cases fully loaded the car?

16 Mrs Threadneedle has enough money to buy 14.7 metres of dress material costing £2·25 per metre. What length of a different material costing £1·75 per metre could she have bought instead?

Part 6 A mixture of direct and inverse proportion

1 3 men are employed to build a wall in 15 hours. How long would 5 men take?

2 If 6 ballpoint pens cost 99 pence, how much will 5 of them cost?

3 In 3 hours I drive 156 miles on the motorway at a steady speed. How far have I travelled in 2 hours?

4 An electric light uses 4 units of electricity in 120 minutes. If 9 units have been used, how long has it been switched on?

5 A mountaineering party takes enough supplies to last 12 men for 6 days. How long would the same supplies last 8 men?

6 My three dogs eat a bag of biscuits every 16 days. If I get another dog, how long will a bag of biscuits last?

7 A cyclist averaging a speed of 24 km/h takes 42 minutes to get to work. How long would he take at 18 km/h?

8 Mr Whey can milk 12 cows by hand in 64 minutes. How long would it take him to milk 15 cows?

9 A machine stamps out 650 plastic trays every 60 seconds. How long does it take to stamp out 390 trays?

10 Bert earns £96 each week for a period of 35 weeks. How long does it take Fred to earn the same amount, if he gets £105 each week?

11 Averaging a speed of 72 km/h, Mr Beckwith takes 4 hours on a journey. How long would it have taken at a speed of 64 km/h?

12 If £12 can be changed for 58.2 Deutschmark, how many DM will you get for £8·60?

13 If 15 francs can be changed for £1.80, how much British money will you get for 13.5 francs?

14 A school has two playing fields. It takes 2 hours to mow the first field with an area of 5.6 hectares. Find the area of the other field, if it takes $1\frac{1}{2}$ hours to mow?

15 If a 15-kg box of peaches cost £4·80, how much will a 24-kg box of peaches cost?

The unitary method

16 If a 32-kg sack of potatoes costs £1·92, what is the mass of a sack costing £3·24?

17 If a gas-ring boils a pan of water in $7\frac{1}{2}$ minutes, how long will two gas-rings take to boil two similar pans of water?

18 A car-park with lines 2.5 metres apart, can hold 360 cars. How many cars would it hold, if the lines were 3 metres apart?

19 It takes 144 street lights 22 metres apart to light a stretch of road. How many lights would be needed, if they were 18 metres apart?

20 A man uses 175 beech saplings to plant a hedge 35 metres long. If he plants another hedge 24 metres long with beech the same distance apart, how many saplings will he use?

Part 7 A mixture of harder problems

1 4 men take 3 days to dig a trench 48 metres long. How long will a second trench be, if it takes 6 men 5 days to dig?

2 3 men build a wall 25 metres long in 5 days. How many days would it take 2 men to build a similar wall 30 metres long?

3 6 people spend 3 days in a hotel and are charged £180. How much would 5 people pay for 4 days in the same hotel?

4 7 businessmen pay £294 to stay 2 days in a hotel. How many days could 5 businessmen stay in the same hotel for £420?

5 12 bags of oats will feed 3 horses for 8 days. How long will 10 bags feed 4 horses with similar appetites?

6 In 6 days 5 horses eat 7 bags of oats. How many bags will be needed to feed 9 horses for 15 days?

7 Mr Bright uses 3 tubes of toothpaste in 9 weeks, if he brushes his teeth twice each day. How many times a day would he brush his teeth, if he used 4 tubes in 8 weeks?

8 Miss Ringlet washes her hair 3 times each week and in 4 weeks uses 5 bottles of shampoo. How many bottles will she use in 6 weeks, if she washes her hair twice a week?

9 In 10 minutes 3 machines punch 480 holes into sheets of metal. How long would 2 machines take to punch 512 holes?

10 4 machines take 12 minutes to staple 216 sheets of paper. How many sheets can 5 machines staple in 10 minutes?

11 It costs £120 to transport 4 tonnes of produce a distance of 140 km by road. How far can you transport 5 tonnes for a payment of £75?

12 15 tonnes of sand are taken 25 km by lorry at a cost of £60. How much would it cost to transport 12 tonnes 35 km?

13 8 women pack 1344 boxes of biscuits in 2 hours. How long would it take 6 women to pack 630 boxes?

14 In 6 hours 15 women pack 186 boxes of biscuits. How many boxes will 12 women pack in 5 hours?

15 Jack Barnes has 3 lorries which use 960 litres of fuel in 2 weeks. How much fuel would 5 lorries use in 3 weeks at the same rate?

16 Another firm has 7 lorries which use 350 litres of fuel in 4 weeks. How long would it take 5 of these lorries to use 625 litres of fuel?

The unitary method

17 3 maths teachers can invent 20 problems in 12 minutes. How long would it take 2 teachers to invent 15 problems?

18 If in 40 minutes 5 maths teachers can mark 120 exam scripts, how many scripts can 7 teachers mark in 65 minutes?

19 Water from 5 similar taps will fill a 1050-litre tank in 75 minutes. How long would water from 7 of these taps take to fill a 2450-litre tank?

20 How long will it take to fill a 504-litre tank with water from 4 taps, if water from 5 similar taps take 35 minutes to fill a 1225-litre tank?

Tables and graphs for proportion

Part 1

Are these pairs of variables **directly** or **inversely** proportional?

As one variable increases, does the other *increase* or *decrease*?

1. Number of articles bought — Amount of money required
2. Distance travelled at a steady speed — Time taken for the journey
3. Length of a ladder — Number of rungs
4. Number of people sharing a Pools win — Amount of money each receives
5. Volume of a certain liquid — Mass of the same liquid
6. Number of pieces cut from a cake — Mass of each piece
7. Number of tractors ploughing a field — Time taken to complete the task
8. Average speed during a journey — Time taken for the journey
9. Petrol used in a car — Distance travelled by the car
10. Size of tyres on a vehicle — Amount of wear on the tyre
11. Number of children sharing a packet of sweets — Number of sweets they each receive
12. Depth of water in a rain-gauge — Length of time it rains
13. Radius of a circle — Circumference of the circle
14. Radius of a bicycle wheel — Number of turns it makes in a given distance
15. Number of men employed — Length of wall they can build in a day
16. Volume of a gas — Pressure of the gas (at constant temperature)

Write four more pairs of variables and say whether they are directly or inversely proportional.

Part 2 Direct proportion

The variables in these tables are **directly** proportional. Copy and complete each table.

1

Sweets (gram)	1	2	3	4	5	10	15	20
Cost (p)	2							

2

Distance cycled (km)	1	2	4	8	12	16	20
Time taken (min)	4						

3

Temperature rise (°C)	1	2	3	6	9	12	20
Time taken (min)	3						

4

Sheet metal (m^2)	$\frac{1}{2}$	1	2	3	4	6	10	20
Mass (kg)	1							

5

Petrol used (litres)	1	$1\frac{1}{2}$	2	$2\frac{1}{2}$	3	4	5	10
Distance travelled (km)	20							

Tables and graphs for proportion

6	Loaves of bread	1	3	4	6	10	12	20
	Cost (p)	40						

7	Train fare (p)	5						
	Length of trip (km)	1	10	20	50	75	100	200

8	American dollar	1.50					
	Pounds sterling (£)	1	2	3	5	10	20

9	Rate per hour (£)	1·70					
	Hours worked	1	2	5	10	12	15

10	Cost of carpet (£)	4·40					
	Length of carpet (m)	1	2	5	10	12	15

These pairs of variables are also directly proportional. Copy and complete the tables. Where possible, start by finding the number paired with 1.

11	Pounds sterling (£)	1	2	3	4	5	10	20
	Deutschmark (DM)		12					

12	Pounds sterling (£)	1	5	10	15	20	50	100
	French francs (F)		40					

13	Time taken (hours)	1	2	5	8	10	12	15
	Distance travelled (miles)		100					

14	Rolls of wallpaper	1	2	3	4	5	10	15
	Price paid (£)					20		

15	Distance motored (miles)		32				
	Petrol used (litres)	1	4	8	10	12	20

16	Volume of concrete (m³)	$\frac{1}{4}$	$\frac{1}{2}$	1	2	5	10
	Mass of concrete (kg)			160			

17	Money invested (£)	50	100	200	300	500	1000
	Interest gained (£)			30			

18	Units of electricity used	10	50	100	150	200	500
	Cost of electricity (£)					24	

19	Area of garden (m²)	25	50	100	200	250	300
	Cost of turfing garden (£)				120		

20	Depth of water in rain-gauge (mm)	5	10	12	15	18	20
	Duration of rain (min)				45		

Tables and graphs for proportion

21 In each of these tables, y is **directly** proportional to x or, in symbols, $y \propto x$.

Copy and complete each table.

Plot a graph of the values in the table.

Find the equation connecting x and y.

a

x	1	2	3	4	5	6
y		6				

b

x	1	2	3	4	5	6
y			6			

c

x	1	2	3	4	5	6
y	$\frac{1}{2}$					

d

x	1	2	3	4	5
y		8			

Choose new scales for the graphs of these tables.

e

x	1	5	10	15	20	25
y		10				

f

x	1	5	10	15	20	25
y			30			

g

x	1	20	40	60	80	100
y			30			

h

x	1	20	40	60	80	100
y						25

22 For each of these tables, write the equation connecting x and y.

a

x	1	2	3	4	5	6
y	4	8	12	16	20	24

b

x	1	2	3	4	5	6
y	6	12	18	24	30	36

c

x	1	5	10	15	20
y	5	25	50	75	100

d

x	1	5	10	15	20
y	$\frac{1}{2}$	$2\frac{1}{2}$	5	$7\frac{1}{2}$	10

e

x	1	20	40	60	80
y	3	60	120	180	240

f

x	1	20	40	60	80
y	2	40	80	120	160

g

x	$\frac{1}{4}$	$\frac{1}{2}$	1	2	4	8
y	$1\frac{1}{2}$	3	6	12	24	48

h

x	6	8	12	15	18
y	24	32	48	60	72

23 An experiment is done to find if there is a connection between the extension (y cm) of a coiled spring and the mass (x grams) which is hung on its end.

The results are shown in this table.

x	10	20	30	40	50	60
y	2.0	4.5	5.5	8.2	10.5	11.6

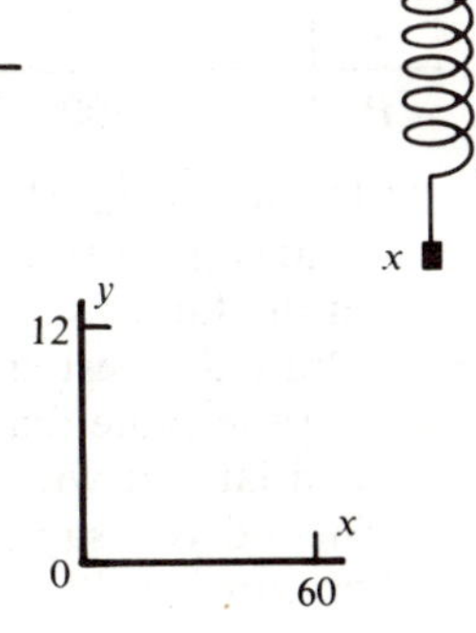

a Draw and label the x-axis from 0 to 60 and the y-axis from 0 to 12. Plot the six points given in this table.

b Draw the best straight line through the points.

c On the same diagram, draw the line $y = \frac{1}{5}x$ and label it with its equation.

d Would you say that y was directly proportional to x?

Tables and graphs for proportion

24 In an electrical experiment the voltage (V volts) across a resistor was measured for different values of current (I amps) flowing through the resistor.

The results are shown in this table.

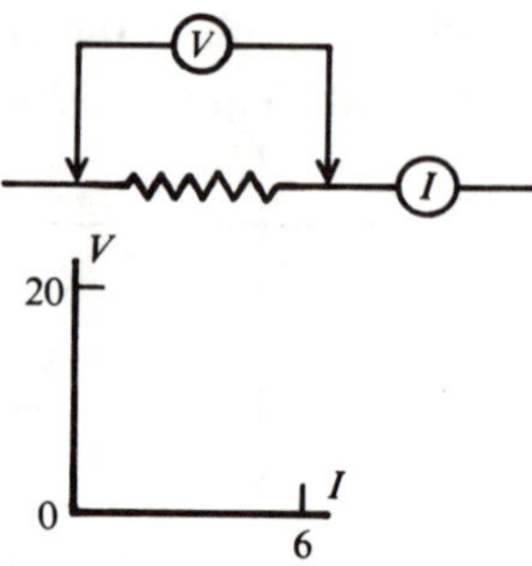

I	1	2	3	4	5	6
V	2	7	9	11	16	18

a Draw and label the I-axis from 0 to 6 and the V-axis from 0 to 20. Plot the six points given in the table.

b Draw the best straight line through the points.

c On the same diagram, draw the line $V = 3I$ and label it with its equation.

d Would you say that V was directly proportional to I?

25 In a chemistry experiment, a quantity of gas is kept at a constant pressure, and its volume (V cm^3) is measured as its temperature (T °K) is increased.

The results are shown in this table.

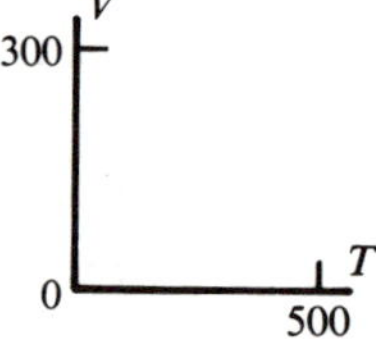

T	100	200	300	400	500
V	60	110	160	180	240

a Draw and label the T-axis from 0 to 500 and the V-axis from 0 to 300. Plot the five points given in the table.

b Draw the best straight line through the points.

c On the same diagram, draw the line $V = \frac{1}{2}T$ and label it with its equation.

d Would you say that V was directly proportional to T?

26 The power (P watts) absorbed by a resistor with a constant voltage across it, changes with the current (I amps) flowing through the resistor.

Results are taken for different currents and are shown in this table.

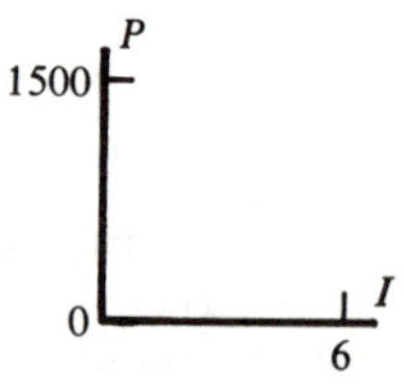

I	1	2	3	4	5	6
P	230	500	730	950	1140	1460

a Draw and label the I-axis from 0 to 6 and the P-axis from 0 to 1500. Plot the six points given in the table.

b Draw the best straight line through the points.

c On the same diagram, draw the line $P = 240I$ and label it with its equation.

d Would you say that P and I were directly proportional?

Tables and graphs for proportion

27 To summarise this part, copy this table.

Statement	Symbols	Equation	Sketch graph
y is directly proportional to x	$y \propto x$	$y = kx$	

Part 3 Inverse proportion

The variables in these tables are **inversely** proportional.

Copy and complete each table.

1

Time taken (h)	1	2	3	4	5	10
Average speed (km/h)	60					

2

Number of men employed	1	2	3	5	10	12
Time taken to do job (h)	120					

3

Volume of gas (litres)	1	2	4	5	8	10	20
Pressure of gas (N/m^2)	40						

4

Number of children sharing a prize	1	2	4	5	6	8	10
Amount each receives (£)	240						

5

Number of taps filling tank	1	2	3	4	6	8	9
Time taken (min)	72						

6

Number of helpings per packet	1	2	3	6	9	12	18
Mass of each helping (grams)	36						

7

Number of apples per tray	10	15	20	30	40	50
Number of trays needed	120					

8

Number of pieces in a cake	4	6	8	12	24
Mass of each piece (grams)	120				

9

Number of pumps emptying a pond	1	2	3	5	6	9	10
Time taken (h)	45						

10

Number of revs per trip	120					
Radius of wheel (cm)	10	12	15	30	40	60

Tables and graphs for proportion

11

x	1	2	4	6	8	12
y		12				

16

x	1	6	12	18	24	36
y				4		

12

x	1	3	6	10	15	30
y		20				

17

x	$\frac{1}{4}$	$\frac{1}{2}$	1	4	8	12
y					2	

13

x	1	5	12	15	20	30
y		24				

18

x	$\frac{1}{4}$	$\frac{1}{2}$	1	2	4	6	9
y				18			

14

x	1	3	6	8	9	12
y			12			

19

x	0.1	0.5	1.0	1.5	2.0
y			24		

15

x	1	2	7	8	14	16
y				4		

20

x	0.1	0.2	0.5	1.0	1.5	2.0
y			4.5			

21 In each of these tables y is **inversely** proportional to x, or, in symbols, $y \propto \dfrac{1}{x}$.
Copy and complete each table.
Plot a graph of the values in the table.
Find the equation connecting x and y.

a

x	1	2	3	4	6	12
y			4			

b

x	1	2	4	8	16
y			2		

c

x	2	4	6	8	12
y			3		

d

x	4	6	8	9	12
y			$4\frac{1}{2}$		

e

x	3	4	6	8	12	16
y				4		

f

x	2	4	8	16	24
y		8			

g

x	4	5	6	8	10	12
y					5	

h

x	4	6	7	8	12	14
y	21					

Each of the curves you have plotted is called a **hyperbola**.

22 For each table, write the equation connecting x and y.

a

x	1	2	3	4	5
y	20	10	$6\frac{2}{3}$	5	4

b

x	1	2	3	4	5
y	40	20	$13\frac{1}{3}$	10	8

c

x	2	4	5	8	10
y	40	20	16	10	8

d

x	2	3	5	6	9	10
y	$22\frac{1}{2}$	15	9	$7\frac{1}{2}$	5	$4\frac{1}{2}$

Tables and graphs for proportion

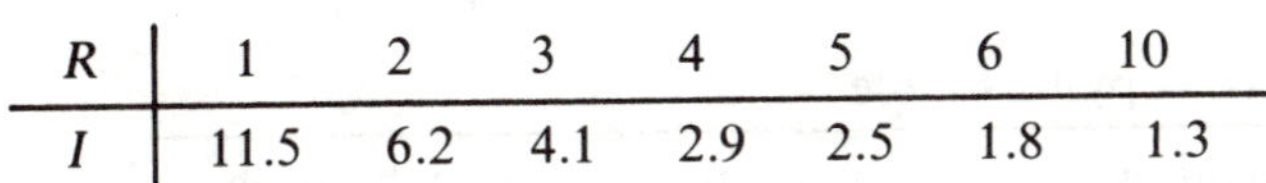

e

x	2	5	10	20	25
y	50	20	10	5	4

f

x	3	5	10	15	25
y	25	15	$7\frac{1}{2}$	5	3

g

x	4	8	16	20	80
y	40	20	10	8	2

h

x	4	8	9	16	24
y	36	18	16	9	6

23 An experiment is done to see if there is a connection between the electrical resistance (R ohms) of a circuit connected across a given battery, and the current (I amps) which the circuit carries.

The results are shown in this table.

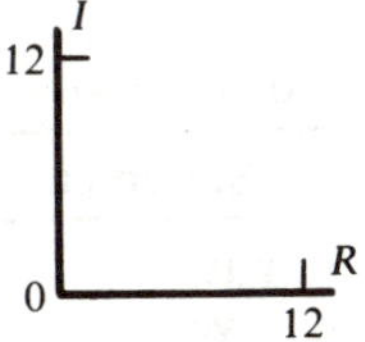

R	1	2	3	4	5	6	10
I	11.5	6.2	4.1	2.9	2.5	1.8	1.3

a Draw and label both axes from 0 to 12.

Plot the seven points given in this table.

b Draw the best smooth curve through these points.

c On the same diagram, draw the hyperbola $I = \dfrac{12}{R}$ and label it with its equation.

d Would you say that I was inversely proportional to R?

24 In another electrical experiment, the voltage (V volts) across a heating element and the current (I amps) flowing through it are adjusted so that the heating element always uses the same amount of power.

The results are shown in this table.

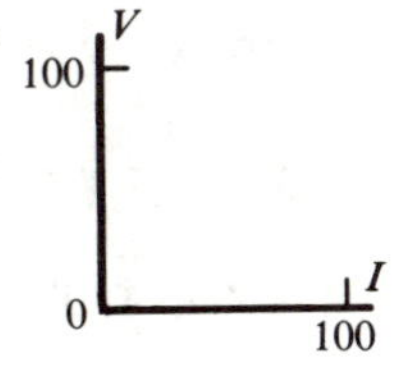

V	4	5	8	10	12	16	20
I	26	19	15	9	8	6	5

a Draw and label both axes from 0 to 100.

Plot the seven points given in this table.

b Draw the best smooth curve through the points.

c On the same diagram, draw the hyperbola $V = \dfrac{100}{I}$ and label it with its equation.

d Would you say that the voltage and current are inversely proportional?

Tables and graphs for proportion

25 The volume (V litres) of a gas is allowed to alter so that its pressure (P kN/m^2) changes, but its temperature stays constant. The pressure and volume are measured several times and the results shown here.

V	7.6	4.9	3.5	3.1	2.6	1.7	1.4
P	2	3	4	5	6	8	10

a Draw and label both axes from 0 to 10.
Plot the seven points given in this table.

b Draw the best smooth curve through the points.

c On the same diagram, draw the hyperbola $P = \dfrac{15}{V}$ and label it with its equation.

d Would you agree with Robert Boyle who in the year 1662 said that the volume of a gas was inversely proportional to its pressure (at constant temperature)?

26 To summarise this part, copy this table.

Statement	Symbols	Equation	Sketch graph
y is inversely proportional to x	$y \propto \dfrac{1}{x}$	$y = \dfrac{k}{x}$	

Part 4 Different types of proportion

1 For each of these, find
(i) the constant of proportionality   (ii) the equation connecting x and y.

a $y \propto x$; $y = 8$ when $x = 2$

b $y \propto x$; $y = 30$ when $x = 6$

c $y \propto x$; $y = 28$ when $x = 4$

d $y \propto \dfrac{1}{x}$; $y = 5$ when $x = 2$

e $y \propto \dfrac{1}{x}$; $y = 3$ when $x = 6$

f $y \propto \dfrac{1}{x}$; $y = 4$ when $x = 4$

g $y \propto x^2$; $y = 45$ when $x = 3$

h $y \propto x^2$; $y = 100$ when $x = 5$

i $y \propto x^2$; $y = 32$ when $x = 4$

j $y \propto \dfrac{1}{x^2}$; $y = 2$ when $x = 3$

k $y \propto \dfrac{1}{x^2}$; $y = 3$ when $x = 5$

l $y \propto \dfrac{1}{x^2}$; $y = 1$ when $x = 4$

m $y \propto \sqrt{x}$; $y = 12$ when $x = 9$

n $y \propto \sqrt{x}$; $y = 30$ when $x = 36$

o $y \propto \sqrt{x}$; $y = 8$ when $x = 4$

p $y \propto x^3$; $y = 24$ when $x = 2$

q $y \propto x^3$; $y = 54$ when $x = 3$

r $y \propto x^3$; $y = 7$ when $x = 1$

s $y \propto \dfrac{1}{\sqrt{x}}$; $y = 4$ when $x = 9$

t $y \propto \dfrac{1}{\sqrt{x}}$; $y = 5$ when $x = 4$

u $y \propto \dfrac{1}{x^3}$; $y = 1$ when $x = 3$

v $y \propto \dfrac{1}{x^3}$; $y = 3$ when $x = 2$

Tables and graphs for proportion

2 For each of these

(i) find the constant of proportionality
(ii) find the equation connecting x and y
(iii) copy and complete the table.

a $y \propto x$

x	1	2	3	4	5
y		8			

b $y \propto x$

x	2	4	6	8	10
y		12			

c $y \propto x$

x	$\frac{1}{2}$	1	2	5	15
y			10		

d $y \propto \dfrac{1}{x}$

x	1	2	3	4	5
y			4		

e $y \propto \dfrac{1}{x}$

x	2	4	6	8	10	12
y			8			

f $y \propto \dfrac{1}{x}$

x	2	4	8	10	16
y		8			

g $y \propto x^2$

x	1	2	3	4	5
y		12			

h $y \propto x^2$

x	1	2	4	5	10
y		20			

i $y \propto x^2$

x	2	3	4	5	6
y			8		

j $y \propto \dfrac{1}{x^2}$

x	1	2	3	4	6	12
y				4		

k $y \propto \dfrac{1}{x^2}$

x	1	2	4	8	10
y		4			

l $y \propto \sqrt{x}$

x	1	4	9	16	25
y			12		

m $y \propto \sqrt{x}$

x	4	16	36	64	100
y			40		

n $y \propto x^3$

x	1	2	3	4	5
y		32			

o $y \propto \dfrac{1}{\sqrt{x}}$

x	1	4	9	16	25
y			20		

p $y \propto \dfrac{1}{x^3}$

x	1	2	4	8	10
y			8		

Tables and graphs for proportion

In each of these, find the equation relating x and y, and then use it to answer the problem.

3 If $y \propto x$ and $y = 12$ when $x = 2$, find y when $x = 5$.

4 If $y \propto x$ and $y = 8$ when $x = 4$, find y when $x = 7$.

5 If $y \propto x$ and $y = 15$ when $x = 3$, find y when $x = 6$.

6 If $y \propto \dfrac{1}{x}$ and $y = 3$ when $x = 8$, find y when $x = 2$.

7 If $y \propto \dfrac{1}{x}$ and $y = 9$ when $x = 4$, find y when $x = 3$.

8 If $y \propto \dfrac{1}{x}$ and $y = 8$ when $x = 9$, find y when $x = 4$.

9 Given that $y \propto x^2$ and $y = 18$ when $x = 3$, find y when $x = 4$.

10 Given that $y \propto x^2$ and $y = 5$ when $x = 1$, find y when $x = 6$.

11 Given that $y \propto x^2$ and $y = 50$ when $x = 10$, find y when $x = 8$.

12 Find y when $x = 6$, given that $y \propto \dfrac{1}{x^2}$ and $y = 4$ when $x = 3$.

13 Find y when $x = 4$, given that $y \propto \dfrac{1}{x^2}$ and $y = 2$ when $x = 8$.

14 Find y when $x = 2$, given that $y \propto \dfrac{1}{x^2}$ and $y = 4$ when $x = 5$.

15 If $y \propto \sqrt{x}$ and $y = 40$ when $x = 100$, find y when $x = 25$.

16 If $y \propto \sqrt{x}$ and $y = 18$ when $x = 9$, find y when $x = \frac{1}{4}$.

17 Find y when $x = 3$, if $y \propto x^3$ and $y = 48$ when $x = 2$.

18 Find y when $x = 4$, if $y \propto x^3$ and $y = 250$ when $x = 5$.

19 Given that $y \propto \dfrac{1}{\sqrt{x}}$ and $y = 3$ when $x = 64$, find y when $x = 9$.

20 Given that $y \propto \dfrac{1}{\sqrt{x}}$ and $y = 8$ when $x = 36$, find y when $x = 16$.

21 The time t for a pendulum to make one full swing is proportional to the square root of its length l. If $t = 10$ when $l = 25$, find
    a an equation relating t and l **b** the value of t when $l = 36$.

22 The volume V of a sphere is proportional to the cube of its radius r. If $V = 32$ when $r = 2$, find
    a an equation relating V and r **b** the value of V when $r = 3$.

23 The braking distance d of a car is proportional to the square of its speed v. If $d = 128$ when $v = 8$, find
    a an equation relating d and v **b** the value of d when $v = 6$.

24 The distance d which an object falls from rest is proportional to the square of the time t taken to fall. If $d = 100$ when $t = 5$, find
    a an equation relating d and t **b** the value of d when $t = 9$.

25 The greatest speed v at which a train can go round a curve, is proportional to the square root of the radius r of the curve. If $v = 40$ when $r = 64$, find
    a an equation relating v and r **b** the value of v when $r = 100$.

Tables and graphs for proportion

26 The gravitational force F between two masses is inversely proportional to the square of the distance d between them. If $F = 5$ when $d = 4$, find
   a an equation relating F and d b the value of F when $d = 2$.

27 The intensity I of light on a surface varies inversely as the square of the distance d of the light source from the surface. If $I = 8$ when $d = 3$, find
   a an equation relating I and d b the value of I when $d = 2$.

28 The energy E of a moving object varies as the square of its velocity v. If $E = 80$ when $v = 4$, find
   a an equation relating E and v b the value of E when $v = 6$.

29 The energy E absorbed by a resistor varies as the square of the current I flowing through it. If $E = 144$ when $I = 6$, find
   a an equation relating E and I b the value of E when $I = 3$.

30 The pressure P of a gas at constant temperature varies inversely as its volume V. If $P = 5$ when $V = 6$, find
   a an equation relating P and V b the value of P when $V = 4$.

31 The frequency f of the note made on an organ varies inversely as the length l of the pipe. If $f = 150$ when $l = 2$, find
   a an equation relating f and l b the value of f when $l = 3$.

32 The time t taken for a journey varies inversely as the average speed v during the journey. If $t = 3$ when $v = 60$, find
   a an equation relating t and v b the value of t when $v = 40$.

33 If $y \propto x^2$ and $y = 36$ when $x = 2$, find y when $x = 5$.

34 If $y \propto x^2$ and $y = 3$ when $x = \frac{1}{2}$, find y when $x = 2$.

35 Given that $y \propto \sqrt{x}$ and $y = 5$ when $x = \frac{1}{4}$, find y when $x = \frac{1}{9}$.

36 If y varies as the square root of x and $y = 1$ when $x = 4$, find y when $x = \frac{1}{4}$.

37 If y is proportional to the cube of x and $y = 2$ when $x = \frac{1}{2}$, find y when $x = 2$.

38 Find y when $x = 2$, if $y \propto x^3$ and $y = 9$ when $x = 3$.

39 If $y \propto \dfrac{1}{x}$ and $y = 6$ when $x = 3$, find y when $x = \frac{1}{2}$.

40 If y is inversely proportional to x, find y when $x = \frac{1}{3}$, given that $y = 2$ when $x = 4$.

41 If $y \propto \dfrac{1}{x^2}$ and $y = 2$ when $x = 3$, find y when $x = 2$.

42 If y varies inversely as the square of x, and $y = 3$ when $x = 4$, find y when $x = 3$.

43 $A \propto r^2$ and $A = 12$ when $r = \frac{1}{2}$. Find A when $r = 2$.

44 $p \propto \dfrac{1}{v}$ and $p = 4$ when $v = 5$. Find v when $p = 2$.

45 $m \propto n^2$ and $m = 75$ when $n = 5$. Find n when $m = 48$.

46 $F \propto \dfrac{1}{d^2}$ and $F = 1$ when $d = 6$. Find d when $F = 4$.

47 $L \propto \sqrt{A}$ and $L = 15$ when $A = 9$. Find A when $L = 30$.

48 $e \propto i^2$ and $e = 5$ when $i = \frac{1}{2}$. Find i when $e = 180$.

49 $x \propto \dfrac{1}{\sqrt{y}}$ and $x = 9$ when $y = 64$. Find y when $x = 36$.

50 $p \propto \sqrt[3]{q}$ and $p = 10$ when $q = 8$. Find q when $p = 20$.

Indices and Computation

Prime numbers

1 Copy this 10 × 10 square grid as shown, with number 1 blacked out.

Imagine the grid as a sieve holding all the whole numbers from 2 to 100.

This square grid is known as Eratosthenes' sieve.

On the first shake of the sieve, the number 2 is caught, but all other numbers which 2 divides into exactly fall through. Cross out all these other numbers.

On the second shake of the sieve, the number 3 is caught, but all other numbers which 3 divides into exactly fall through. Cross out all these other numbers.

■	2	3	4	5	6	7	8	9	10
11	12	13	14	15	16	17	18	19	20
21	22	23	24	25	26	27	28	29	30
31	32	33	34	35	36	37	38	39	40
41	42	43	44	45	46	47	48	49	50
51	52	53	54	55	56	57	58	59	60
61	62	63	64	65	66	67	68	69	70
71	72	73	74	75	76	77	78	79	80
81	82	83	84	85	86	87	88	89	90
91	92	93	94	95	96	97	98	99	100

On the third shake of the sieve, the number 5 is caught, but all other numbers which 5 divides into exactly fall through. Cross out all these other numbers.

On the next shake of the sieve, the number 7 is caught. Cross out those numbers which 7 divides into exactly.

Repeat this until no more numbers can be crossed out. You are left with the prime numbers. **A prime number** is a number which no number (except 1 and itself) divides into exactly.

Write a list of the prime numbers from 1 to 100.

2 Break down each of these numbers into a row of prime numbers which when multiplied together, will give that number.

Example

```
2 | 720
2 | 360
2 | 180
2 | 90
3 | 45
3 | 15
5 | 5
  | 1
```

So
$720 = 2 \times 2 \times 2 \times 2 \times 3 \times 3 \times 5.$

a 36	b 24	c 120	d 48	e 72	f 200
g 300	h 256	i 144	j 192	k 1000	l 360
m 216	n 162	o 1080	p 729	q 625	r 2700
s 2500	t 7700	u 2940	v 30 030	w 2090	x 25 935
y 7865	z 42 336				

3 When a number is multiplied by itself several times, it can be written using a power (or index). In the Example above we can write $720 = 2^4 \times 3^2 \times 5$.

Write each of these numbers as a row of prime numbers multiplied together, and then write the answer using powers (or indices).

a 16	b 27	c 32	d 128	e 2000
f 432	g 81	h 243	i 900	j 9000

4 Go back to question 2 above and write all the answers there using powers.

Powers or indices

1 Simplify these. Do *not* work out their values.

a $\quad 2 \times 2 \times 2 \times 2$ b $\quad 7 \times 7 \times 7$

c $\quad 12 \times 12 \times 12 \times 12 \times 12$ d $\quad 21 \times 21 \times 21 \times 21$

e $\quad 9 \times 9 \times 9 \times 9 \times 9 \times 9$ f $\quad 13 \times 13 \times 13 \times 13 \times 13 \times 13 \times 13 \times 13$

g $\quad \frac{1}{2} \times \frac{1}{2} \times \frac{1}{2} \times \frac{1}{2} \times \frac{1}{2}$ h $\quad 6 \times 6 \times 6 \times 6$

i $\quad 3 \times 3 \times 3$ j $\quad 4 \times 4 \times 4 \times 4 \times 4$

2 Simplify these.

a $\quad 6 \times 6 \times 6 \times 5 \times 5 \times 5 \times 5$ b $\quad 8 \times 8 \times 7 \times 7 \times 7 \times 7$

c $\quad 2 \times 2 \times 2 \times 2 \times 2 \times 9 \times 9$ d $\quad 3 \times 3 \times 3 \times 3 \times 4 \times 4 \times 4$

e $\quad 3 \times 3 \times 3 \times 4 \times 4 \times 4 \times 4$ f $\quad 12 \times 12 \times 12 \times 8 \times 8$

g $\quad 2 \times 6 \times 6 \times 2 \times 2 \times 6 \times 2$ h $\quad 3 \times 3 \times 7 \times 3 \times 3 \times 7 \times 7 \times 7$

i $\quad 5 \times 5 \times 3 \times 3 \times 5 \times 5 \times 3 \times 3$ j $\quad 8 \times 5 \times 6 \times 5 \times 5 \times 6 \times 6 \times 8 \times 5$

3 Write these in full, and then find their values.

a $\ 2^4$ b $\ 3^3$ c $\ 5^2$ d $\ 2^3$ e $\ 10^3$

f $\ 2^5$ g $\ 1^6$ h $\ 4^2$ i $\ 0^5$ j $\ 6^2$

k $\ 2^6$ l $\ 1^4$ m $\ 5^3$ n $\ 4^3$

4 Find the value of each of these. All the answers should be different.

a $\ 2^3$ b $\ 3^2$ c $\ 3 \times 2$ d $\ 3 + 2$

e $\ 5^2$ f $\ 2^5$ g $\ 2 \times 5$ h $\ 2 + 5$

5 Write these in full, and then find their values.

a $\ 2^2 \times 3^2$ b $\ 2^2 \times 5^2$ c $\ 10^3 \times 3^2$ d $\ 1^3 \times 6^2$ e $\ 2^4 \times 3^2$

f $\ 1^4 \times 7^2$ g $\ 0^4 \times 9^2$ h $\ 3^4 \times 2^3$ i $\ 12^2 \times 2^3$ j $\ 3^4 \times 10^3$

k $\ 4^3 \times 5^2$ l $\ 4^3 \times 3^4$ m $\ 6^2 \times 2^6$ n $\ 5^3 \times 3^5$ o $\ 4^4 \times 3^5$

p $\ 2^4 \times 7^2$ q $\ 10^4 \times 3^3$ r $\ 6^3 \times 2^4$ s $\ 8^2 \times 2^5$ t $\ 7^3 \times 4^2$

6 Simplify these, using indices in the answers.
(Do *not* work out their values.)

a $\quad 6^7 \times 6^5$ b $\quad 8^7 \times 8^{12}$

c $\quad 2^4 \times 2^5$ d $\quad 3^2 \times 3^7$

e $\quad 8^9 \times 8^7$ f $\quad 10^2 \times 10^6 \times 10^3$

g $\quad 3^7 \times 3^{12} \times 3^4$ h $\quad 5^5 \times 5^{10} \times 5^4$

i $\quad 7^9 \times 7^6$ j $\quad 18^2 \times 18^7$

k $\quad 9^4 \times 9^5 \times 9^6$ l $\quad 12^7 \times 12^4 \times 12^2$

m $\quad 10^9 \times 10^5 \times 10^2$ n $\quad 1^4 \times 1^6 \times 1^7$

o $\quad 3^2 \times 3^7 \times 5^4 \times 5^5$ p $\quad 6^7 \times 6^8 \times 4^6 \times 4^3$

q $\quad 8^4 \times 8^7 \times 12^2 \times 12^2$ r $\quad 4^7 \times 4^2 \times 5^6 \times 5^4$

s $\quad 10^6 \times 10^{13} \times 10^7 \times 8^7 \times 8^4$ t $\quad 7^4 \times 8^7 \times 8^6 \times 7^9$

u $\quad 3^2 \times 4^7 \times 3^9 \times 4^7$ v $\quad 7^9 \times 7$

w $\quad 2^9 \times 8^4 \times 2^7 \times 8$ x $\quad 4 \times 10^6 \times 4^2 \times 10^2$

y $\quad 1^3 \times 10^2 \times 1^4 \times 10^2$ z $\quad 12^4 \times 21^6 \times 21 \times 12$

Powers or indices

7 Find the value of each lettered index to make the statement true.

a $2^7 \times 2^a = 2^9$ b $3^6 \times 3^b = 3^8$

c $7^5 \times 7^c = 7^6$ d $8^9 \times 8^d = 8^{15}$

e $10^4 \times 10^e = 10^9$ f $4^2 \times 4^5 \times 4^f = 4^{12}$

g $5^7 \times 5^g \times 5^8 = 5^{20}$ h $12^7 \times 12^h \times 12^2 = 12^{10}$

i $9^i \times 9^8 \times 9 = 9^{12}$ j $(\tfrac{1}{2})^4 \times (\tfrac{1}{2})^j = (\tfrac{1}{2})^6$

k $7^2 \times 7^4 \times 7^k = 7^9$ l $(\tfrac{3}{4})^6 \times (\tfrac{3}{4})^3 \times (\tfrac{3}{4})^l = (\tfrac{3}{4})^9$

m $6^m \times 6^m = 6^8$ n $2^n \times 2^n = 2^{10}$

o $6^x \times 6^x = 6^2$ p $4^p \times 4^p \times 4^p = 4^{15}$

q $8^4 \times 8^q \times 8^q = 8^{10}$ r $9^2 \times 9^r \times 9^r = 9^{12}$

s $3^s \times 3^5 \times 3^s = 3^9$

8 Work these, taking care to place the decimal point correctly.

a 2.67×10 b 2.67×10^2 c 1.27×10

d 1.27×10^2 e 1.27×10^3 f 0.462×10^2

g 0.462×10^3 h 7.8×10 i 7.8×10^2

j 7.8×10^3 k 0.679×10^3 l 0.43×10^3

m 0.0148×10^4 n 0.0148×10^2 o 0.0089×10^3

p 0.0089×10^2 q 6.2×10^4 r 1.8×10^6

s 6.2×10^2 t 0.007×10^3 u 0.007×10^5

v 0.4×10^3 w 0.001×10^4 x 0.01×10^6

y 1×10^3 z 1×10^6

9 Find the value of n in each of these.

a $2^n = 8$ b $3^n = 9$ c $10^n = 1000$ d $5^n = 25$

e $6^n = 36$ f $2^n = 32$ g $5^n = 125$ h $4^n = 64$

i $3^n = 81$ j $2^n = 128$ k $4^n = 256$ l $3^n = 243$

10 Find the area of the square and the volume of the cube when

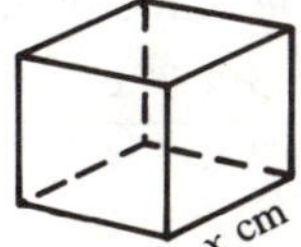

a $x = 3$ b $x = 4$ c $x = 2$ d $x = 5$ e $x = 10$

f $x = 1$ g $x = 6$ h $x = 8$ i $x = 12$ j $x = \tfrac{1}{2}$.

11 Find the values of these.

a $(2^3)^2$ b $(3^2)^2$ c $(5^2)^2$ d $(2^2)^3$ e $(3^3)^2$

f $(4^2)^2$ g $(2^4)^2$ h $(2^2)^4$ i $(1^3)^5$

12 Simplify.

a $a^2 \times a^5$ b $x^4 \times x^7$

c $y^3 \times y^2 \times y^4$ d $m^6 \times m^3 \times r$

e $a \times a^6$ f $b \times b^4$

g $c \times c^2 \times c^6$ h $x \times x^3$

Powers or indices

i $y \times y \times y^7$

j $z \times z^3 \times z$

k $a^4 \times a^6 \times b^3 \times b$

l $m^2 \times m^3 \times n^5 \times n^2$

m $p^2 \times q^4 \times p^2 \times q^3$

n $x \times y^4 \times x^3 \times y$

o $a \times s \times a^3 \times s^2$

p $a^2b^3 \times a^4b^2$

q $m^3n^2 \times mn^4$

r $xy^2 \times x^2y$

s $y^2z^3 \times yz^2$

t $pq^3 \times p^3q$

u $a^2b^2 \times a^3b$

v $m^4n \times mn$

w $ac \times ac$

x $x^2y^2z^2 \times x^3yz$

y $a^2bc \times ab^2c^3$

z $pq^2r \times p^2qr$

13   Simplify.

a $(x^2)^3$ b $(y^4)^2$ c $(z^5)^3$ d $(m^3)^5$ e $(n^4)^2$

f $(x^2y^3)^2$ g $(a^4b^2)^3$ h $(p^3q^5)^2$ i $(m^2n)^5$ j $(yz^3)^4$

k $(2x^2)^3$ l $(3y^4)^2$ m $(2z^3)^4$ n $(5p^4)^3$ o $(6q^3)^2$

p $(2m^4)^5$ q $(3k^5)^3$ r $(10t)^3$ s $(2n)^6$ t $(\tfrac{1}{2}v^6)^2$

14   Simplify.

a $(4x^3)^2$

b $(3y^5)^2$

c $(2m^4)^3$

d $(2n^5)^3$

e $(3x^2)^2$

f $(5p)^3$

g $2x^4 \times (3x^2)^2$

h $5a^4 \times (2a^4)^2$

i $2p^3 \times (3p^4)^3$

j $4q^2 \times (5q^2)^3$

k $(3m^4)^2 \times 5m^4$

l $(2x^3)^4 \times 2x^7$

m $(10c^3)^3 \times 7c^4$

n $(2d^2)^3 \times (3d)^2$

o $(4e^4)^2 \times (5e^3)^2$

15   Find the value of each of these.

a $\dfrac{3 \times 3 \times 3 \times 3 \times 3}{3 \times 3 \times 3}$ b $\dfrac{5 \times 5 \times 5 \times 5 \times 5 \times 5}{5 \times 5 \times 5 \times 5 \times 5}$ c $\dfrac{2 \times 2 \times 2 \times 2 \times 2}{2 \times 2}$

d $\dfrac{2^7}{2^4}$ e $\dfrac{3^7}{3^5}$ f $\dfrac{5^9}{5^8}$

g $\dfrac{8^7}{8^6}$ h $\dfrac{4^5}{4^2}$ i $\dfrac{2^{10}}{2^6}$

j $\dfrac{3^{12}}{3^9}$ k $\dfrac{10^6}{10^3}$ l $\dfrac{9^3}{9}$

m $\dfrac{5^4}{5}$ n $\dfrac{2^5 \times 4^6}{2^2 \times 4^4}$ o $\dfrac{3^8 \times 5^6}{3^7 \times 5^4}$

p $\dfrac{8^6 \times 2^9}{8^5 \times 2^7}$ q $\dfrac{6^5 \times 3^4}{6^3 \times 3^5}$ r $\dfrac{8^9 \times 2^3}{8^7 \times 2^8}$

    $\dfrac{10^6 \times 5^5}{0^3}$ t $\dfrac{3^7 \times 9^4}{3^3 \times 9^5}$

Powers or indices

16 Simplify.

a $\dfrac{x^7}{x^4}$ b $\dfrac{y^8}{y^3}$ c $\dfrac{z^4}{z^2}$ d $\dfrac{m^8n^5}{m^2n^2}$

e $\dfrac{p^7q^6}{pq^2}$ f $\dfrac{a^3b^4}{ab}$ g $\dfrac{y^7z^2}{y^2z^6}$ h $\dfrac{a^4b^3}{ab^7}$

i $\dfrac{6a^8}{3a^4}$ j $\dfrac{12p^7}{3p^2}$ k $\dfrac{3z^4}{9z}$ l $\dfrac{10a^4b^7}{5a^5b}$

m $\dfrac{16m^5n}{8m^3n^3}$ n $\dfrac{2b^2c^4}{6b^7c^2}$ o $\dfrac{3mn^2}{9m^2n}$ p $\dfrac{a^2b^2c^2}{a^3b^3c^3}$

q $\dfrac{2pq}{6p^2q^3}$ r $\dfrac{6a^2b}{4ab}$ s $\dfrac{6m^2n^2}{9mn^3}$ t $\dfrac{12ab^4}{8a^4b}$

u $\dfrac{6st^2u^3}{10st^3u^3}$ v $\dfrac{4p^3qr}{6pqr}$ w $\dfrac{10x^7y}{8x^6y}$ x $\dfrac{12x^2y^3z^4}{4x^3y^3z^3}$

y $\dfrac{3m^4n^2}{6mn^3}$ z $\dfrac{8u^4v^4}{6u^4v^5}$

17 Work these, taking care to place the decimal point correctly.

a $\dfrac{6.7}{10^2}$ b $\dfrac{87.3}{10^2}$ c $\dfrac{21.4}{10^3}$ d $\dfrac{3.6}{10^3}$

e $\dfrac{2}{10^4}$ f $\dfrac{3}{10^2}$ g $\dfrac{0.6}{10^4}$ h $\dfrac{0.72}{10^3}$

i $\dfrac{0.05}{10^2}$ j $\dfrac{128}{10^5}$ k 1.3×10^4 l $\dfrac{1.3}{10^4}$

m 2.8×10^3 n $\dfrac{2.8}{10^3}$ o 0.045×10^4 p 0.25×10^3

q $\dfrac{0.25}{10^3}$ r 0.07×10^4 s 6×10^5 t $\dfrac{6}{10^5}$

u 0.2×10^4 v $\dfrac{0.2}{10^4}$ w $\dfrac{3600}{10^3}$ x $\dfrac{42\,000}{10^5}$

y $\dfrac{1}{10^4}$ z $\dfrac{750}{10^6}$

18 Find the value of each of these.

a $\dfrac{12 \times 10^6}{4 \times 10^2}$ b $\dfrac{18 \times 10^9}{3 \times 10^5}$ c $\dfrac{6 \times 10^5}{2 \times 10^2}$ d $\dfrac{36 \times 10^7}{9 \times 10^6}$

e $\dfrac{21 \times 10^6}{3 \times 10^5}$ f $\dfrac{48 \times 10^{10}}{16 \times 10^9}$ g $\dfrac{42 \times 10^7}{6 \times 10^2}$ h $\dfrac{14 \times 10^3}{7 \times 10^7}$

i $\dfrac{27 \times 10^4}{3 \times 10^9}$ j $\dfrac{8 \times 10^5}{2 \times 10^{10}}$ k $\dfrac{4.8 \times 10^3}{4 \times 10^8}$ l $\dfrac{3.8 \times 10^5}{2 \times 10^9}$

m $\dfrac{6.5 \times 10^6}{5 \times 10^8}$ n $\dfrac{5.6 \times 10^{12}}{2 \times 10^8}$ o $\dfrac{7.2 \times 10^7}{6 \times 10^5}$ p $\dfrac{8.4 \times 10^{10}}{4 \times 10^5}$

q $\dfrac{10^7}{2 \times 10^6}$ r $\dfrac{10^9}{5 \times 10^8}$ s $\dfrac{10^6}{5 \times 10^4}$ t $\dfrac{4 \times 10^{12}}{8 \times 10^{11}}$

Standard form

1. Write these numbers in standard form.
   Example $78\,000 = 7.8 \times 10^4$
a 9 400 000	b 670 000	c 43 500	d 8 530 000
e 3400	f 60 000	g 1000	h 840
i 753	j 2040		

2. Write the populations of these countries in standard form.
a United Kingdom	56 000 000	b Sri Lanka	10 600 000
c USSR	226 000 000	d Iceland	190 000
e China	656 000 000	f The Gambia	330 000
g Vatican City	940		

3. Write these in full.
a 1.24×10^5	b 6.75×10^4	c 1.6×10^6
d 5.03×10^3	e 9.19×10^8	f 1.005×10^5
g 8×10^3	h 2×10^6	i 1×10^5

4. These areas are all measured in km². Write them in full.
a Sahara Desert	8.42×10^6	b Lake Ontario	1.95×10^4
c Australia	7.7×10^6	d Bahrein	5.52×10^2
e San Marino	5.95×10^1	f Earth's surface	5.6×10^8

5. Write these items of information in standard form.
   a The speed of light is 300 000 000 metres per second.
   b The distance from Earth to the moon is 245 000 miles.
   c The speed of sound at ground level is 330 000 cm per second.
   d The diameter of the sun is 865 000 miles.
   e The distance from the Earth to the sun is 92 900 000 miles.
   f The distance from the sun to the outermost planet, Pluto, is 3 670 000 000 miles.

6. Write these in order of size, with the *smallest* first.
a 1.7×10^6	2.3×10^5	8.4×10^3	9×10^2
b 3.7×10^9	1.4×10^5	2×10^7	8×10^3

7. Write these in order of size, with the *largest* first.
a 8.7×10^2	3×10^6	7.2×10^6	10^5	5×10^5
b 1.07×10^5	2×10^4	10^5	9.4×10^4	

8. Find the value of n in these.
a $7.8 \times 10^n = 780\,000$	b $9.7 \times 10^n = 9700$
c $8.25 \times 10^n = 825\,000$	d $1.04 \times 10^n = 1040$
e $3 \times 10^n = 3\,000\,000$	f $7 \times 10^n = 700$
g $2.12 \times 10^n = 212$	h $8.5 \times 10^n = 85$
i $(200)^2 = 4 \times 10^n$	j $(400)^2 = 1.6 \times 10^n$
k $(5000)^2 = 2.5 \times 10^n$	l $(150)^2 = 2.25 \times 10^n$

Standard form

9 Approximate these numbers as requested and then write them in standard form.
  a   628 000 to 2 significant figures   b   97 200 to 2 significant figures
  c   46 600 to 2 significant figures   d   328 400 to 1 significant figure
  e   9 870 000 to 2 significant figures   f   209 450 to 2 significant figures
  g   18 340 to 3 significant figures   h   485 345 000 to 1 significant figure
  i   9172 to 3 significant figures   j   57 334 500 to 2 significant figures

10 Work these and give the answers in standard form.
  a 600×500 b 500×800 c 250×800
  d 4000×500 e 700×4000 f 250×3000
  g 7500×400 h $25 \times 40 \times 600$ i $75 \times 20 \times 400$
  j $4 \times 10^4 \times 2 \times 10^8$ k $3 \times 10^9 \times 3 \times 10^7$
  l $3.5 \times 10^{11} \times 2 \times 10^8$ m $3 \times 10^7 \times 4 \times 10^3$
  n $8 \times 10^9 \times 4 \times 10^4$ o $2.6 \times 10^4 \times 4 \times 10^6$

11 Give answers to these in standard form.
  a $\dfrac{8 \times 10^9}{2 \times 10^5}$ b $\dfrac{9 \times 10^7}{3 \times 10^3}$ c $\dfrac{7 \times 10^6}{2 \times 10^2}$ d $\dfrac{8.2 \times 10^9}{2 \times 10^7}$

  e $\dfrac{6.5 \times 10^{11}}{5 \times 10^6}$ f $\dfrac{4.2 \times 10^8}{7 \times 10^4}$ g $\dfrac{2.1 \times 10^{10}}{3 \times 10^4}$ h $\dfrac{6 \times 10^7}{1.5 \times 10^4}$

  i $\dfrac{10^8}{5 \times 10^3}$ j $\dfrac{10^7}{2 \times 10^4}$

12 a   By writing 60 000 in standard form, find the value of $(60\,000)^2$ in standard form.
  Find the values of these squares and cubes, and give each answer in standard form.
  b $(7000)^2$ c $(5000)^2$ d $(80\,000)^2$ e $(400\,000)^2$
  f $(7000)^3$ g $(5000)^3$ h $(80\,000)^3$ i $(400\,000)^3$
  j $(12\,000)^2$ k $(1100)^2$ l $(130\,000)^2$ m $(12\,000)^3$

13 Find the values of each of these, but do *not* put the answers in standard form.
  a $(6.4 \times 10^3) + (8.2 \times 10^2)$ b $(4.5 \times 10^4) + (3.7 \times 10^3)$
  c $(8 \times 10^5) + (4.5 \times 10^4)$ d $(3.6 \times 10^3) + (8.5 \times 10^2)$
  e $(9.45 \times 10^6) + (2.33 \times 10^5) + (1.5 \times 10^4)$
  f $(8.245 \times 10^5) + (4.22 \times 10^4) + (3 \times 10^3) + (2 \times 10^2)$
  g $(1.234 \times 10^6) + (1.234 \times 10^5) + (1.234 \times 10^4) + (1.234 \times 10^3) + (1 \times 10^2)$
  h $(6.54 \times 10^5) - (3.7 \times 10^4)$ i $(8.57 \times 10^3) - (4.8 \times 10^2)$
  j $(9.07 \times 10^5) - (3 \times 10^3)$ k $(4.5 \times 10^4) - (7 \times 10^2)$

14 Write these volumes in cm^3 in standard form.
  a   12 litres   b   23 litres   c   350 litres   d   540 litres
  e   800 litres   f   3000 litres   g   4500 litres   h   25 000 litres
  i   60 000 litres   j   8.5 litres

Standard form

15 Write these masses in grams in standard form.

a 17 kg	**b** 38 kg	**c** 240 kg	**d** 720 kg
e 900 kg	**f** 2000 kg	**g** 6500 kg	**h** 14 000 kg
i 20 000 kg	**j** 3.5 kg		

16 Write these distances in mm in standard form.

a 3 m	**b** 8 m	**c** 7.5 m	**d** 6.2 m
e 23 m	**f** 47 m	**g** 86.5 m	**h** 126 m
i 237 m	**j** 450 m	**k** 2 km	**l** 4.5 km
m 24 km	**n** 32 km	**o** 160 km	

17 If light travels at a speed of 300 000 000 metres per second, find how many kilometres it travels in one year, giving the answer in standard form to 2 significant figures. (This distance is called **one light year**.)

18 The circumference of the earth is 4×10^4 km. If you walked an average of 25 km per day, how long would it take to walk round this circumference?

Squares

Part 1 Using diagrams and long multiplication

Find the areas of these squares, where

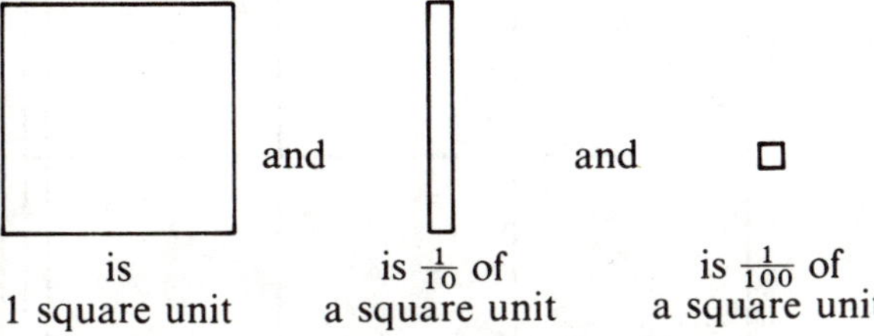

    and   and   □

is
1 square unit   is $\frac{1}{10}$ of
a square unit   is $\frac{1}{100}$ of
a square unit.

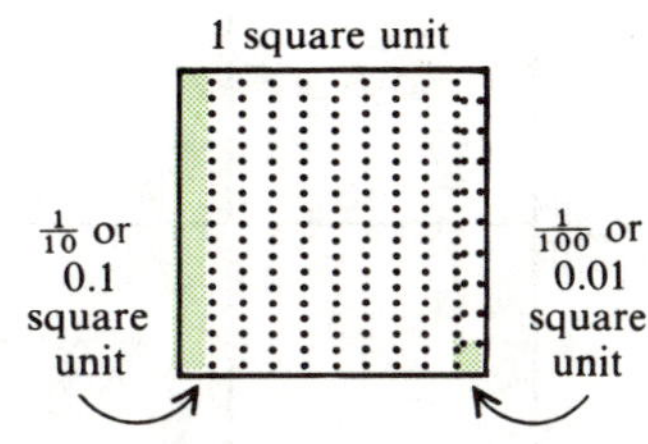

Check each answer by long multiplication.

1

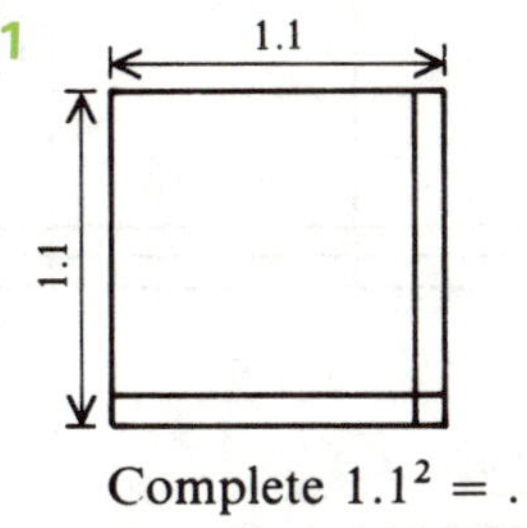

Complete $1.1^2 = \ldots$

2

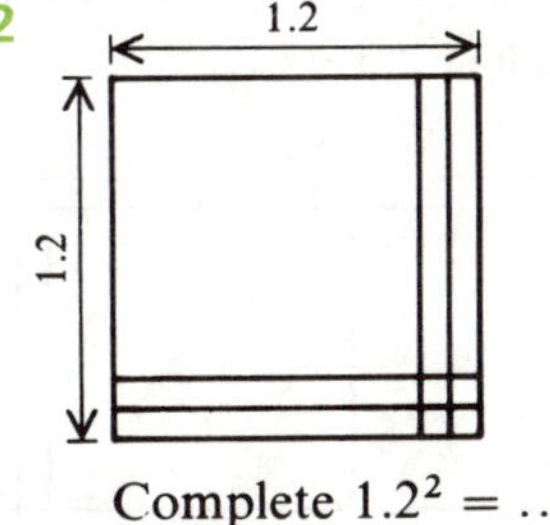

Complete $1.2^2 = \ldots$

3

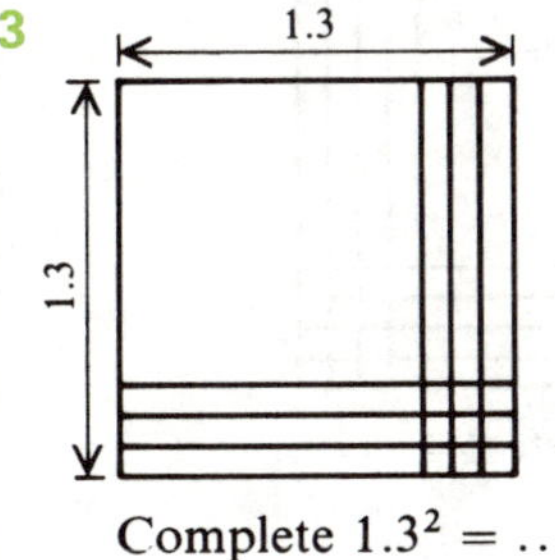

Complete $1.3^2 = \ldots$

4

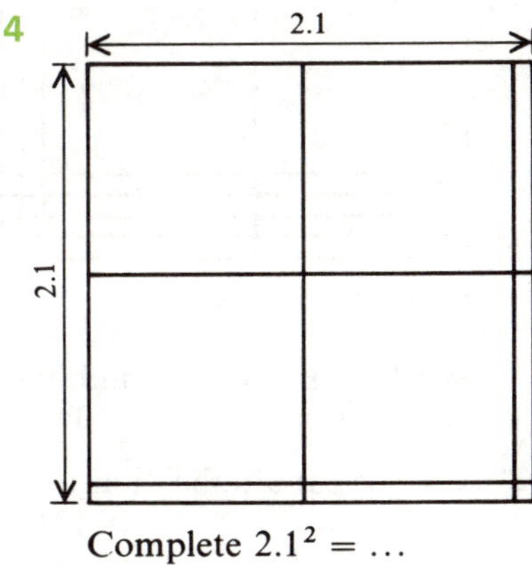

Complete $2.1^2 = \ldots$

5

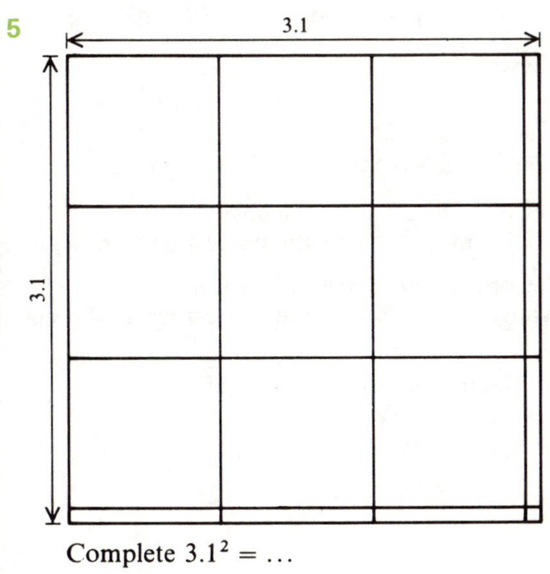

Complete $3.1^2 = \ldots$

6

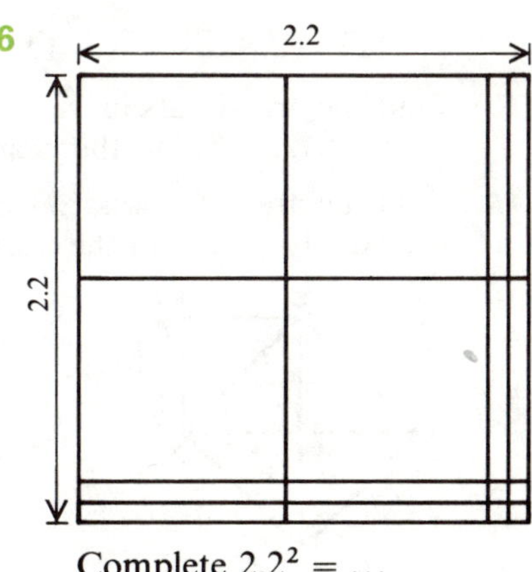

Complete $2.2^2 = \ldots$

Squares

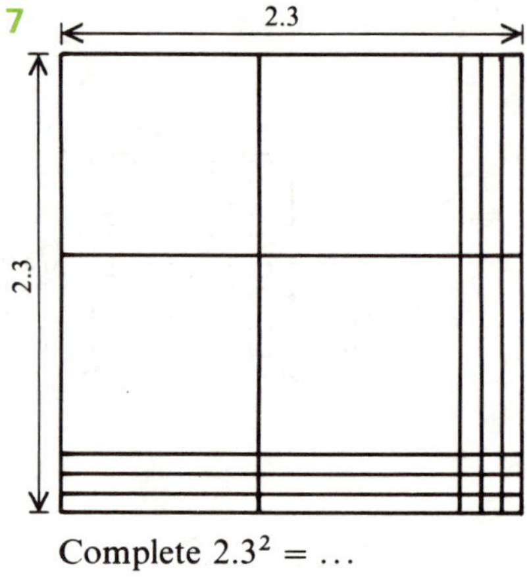

7 2.3 (top), 2.3 (side)

Complete $2.3^2 = \ldots$

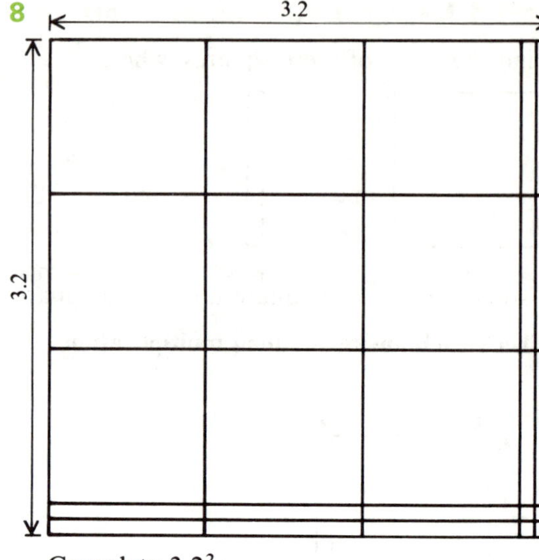

8 3.2 (top), 3.2 (side)

Complete $3.2^2 = \ldots$

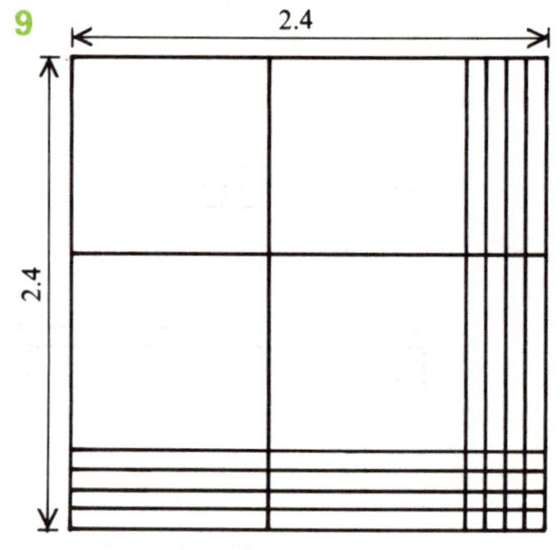

9 2.4 (top), 2.4 (side)

Complete $2.4^2 = \ldots$

10 1.4 (top), 1.4 (side)

Complete $1.4^2 = \ldots$

11 Work these by long multiplication, and **learn by heart**.
10^2 11^2 12^2 13^2 14^2 15^2 16^2 20^2 25^2 30^2 40^2

12 Use long multiplication to work these.

a 20^2	**b** 21^2	**c** 22^2	**d** 22.5^2
e 30^2	**f** 31^2	**g** 32^2	**h** 32.5^2
i 125^2	**j** 125.1^2	**k** 0.25^2	**l** 0.251^2

13 Find the area of a square with sides of 15.6 cm. Give the answer
a exactly **b** to the nearest whole cm^2 **c** to the nearest tenth of a cm^2.

14 Find the area of a square with sides of 27.4 cm. Give the answer
a exactly **b** to the nearest whole cm^2 **c** to the nearest tenth of a cm^2.

15

If $XY = 18$ cm, calculate the area of
a the square $WXYZ$
b the triangle WXZ
c the square $ABXZ$.

Squares

16 The square *ABCD* has sides of 6.5 cm.
A border 2.5 cm wide is placed along two of its
sides to make the square *AXYZ*.

Calculate the area of
a the square *ABCD*
b the square *AXYZ*
c the shaded border.

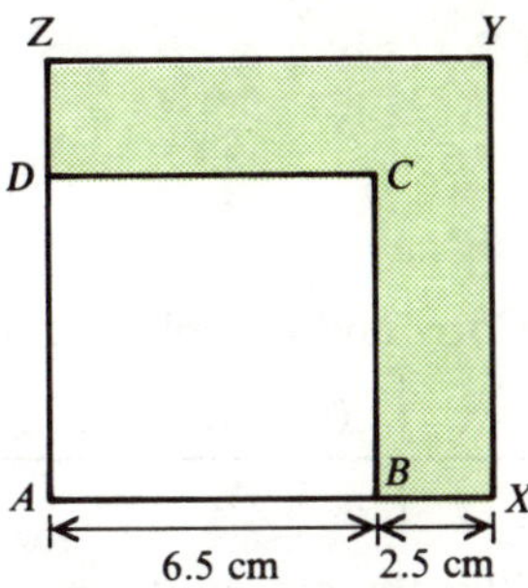

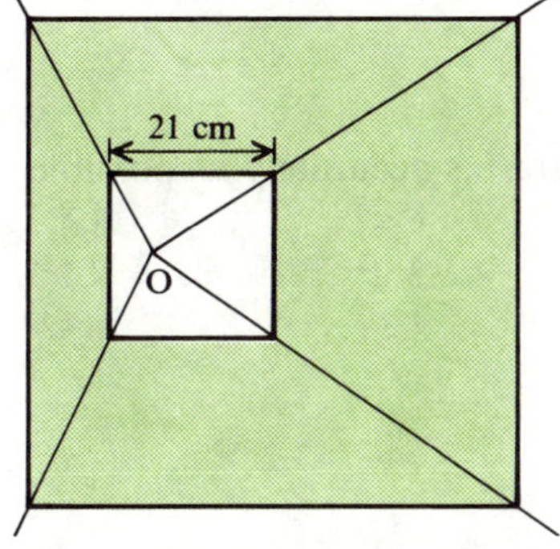

17 A square of side 21 cm is enlarged into a
square with sides three times longer.

Calculate the area of
a the original square
b the enlarged square
c the shaded region.

18 A shape is made from two squares *R* and *S*, and
two triangles *T* and *U*. *R* has sides 16 cm long,
and *S* has sides twice as long as *R*.

Calculate the area of
a square *R*
b square *S*
c triangle *T*
d the whole shape.

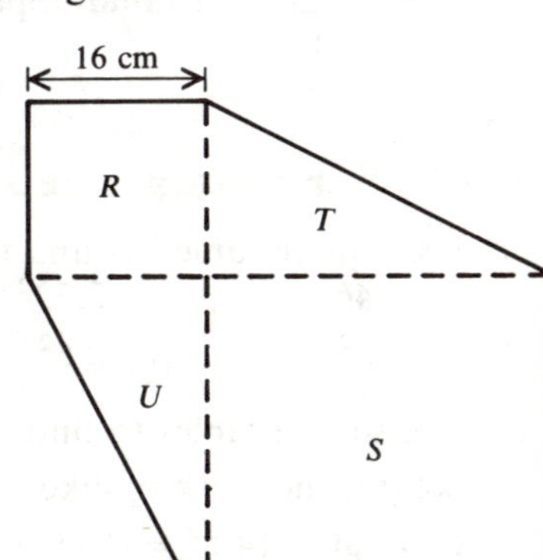

19

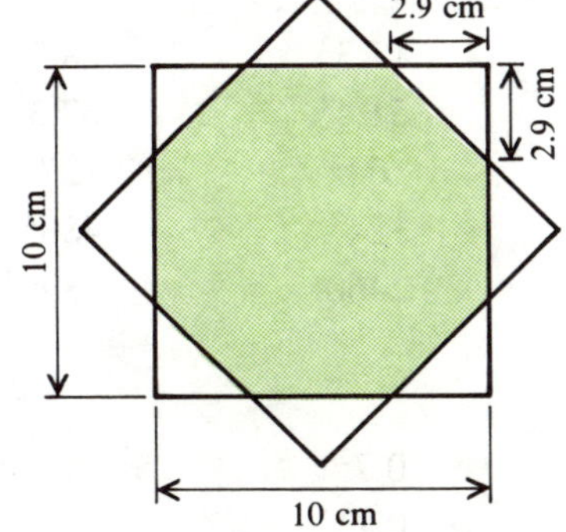

A square of side 10 cm is rotated 45° about
its centre as shown.
a Calculate (to 2 significant figures) the
area of one of the eight right-angled
triangles which are formed.
b Hence calculate the shaded area.

20 Two equal squares *A* and *B* of side 24 cm
are placed in the opposite corners of another,
larger square of side 42 cm.

Calculate the area of
a square *A*
b the larger square
c each shaded square
d the dotted square.

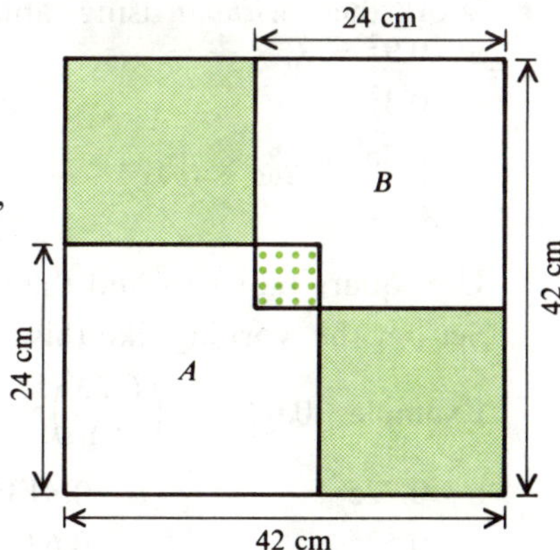

Squares

Part 2 Using a graph

1

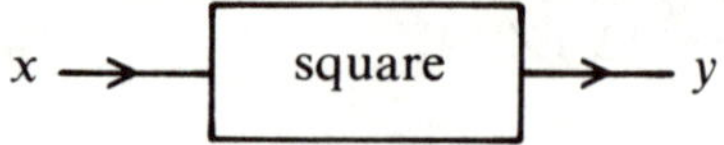

Copy and complete this table for the equation $y = x^2$.

x	0	1	2	3	4	5	6	7	8	9	10
y											

Plot the graph of $y = x^2$ onto squared paper.

Use this graph to write the values of these squares as accurately as possible.

a	5.5^2	**b**	6.5^2	**c**	7.5^2	**d**	8.2^2	**e**	4.8^2
f	6.8^2	**g**	9.3^2	**h**	5.2^2	**i**	3.2^2	**j**	7.1^2
k	2.6^2	**l**	8.8^2	**m**	9.7^2	**n**	1.8^2	**o**	1.5^2

2 Use the graph to find approximate values of
a $(2.5^2)^2$ **b** $(2.8^2)^2$ **c** $(3.2^2)^2$.

Part 3 Using square tables

1 Use square tables to find the values of these.

a	1.46^2	**b**	2.73^2	**c**	3.17^2	**d**	4.47^2	**e**	8.66^2
f	7.5^2	**g**	2.6^2	**h**	7.07^2	**i**	1.3^2	**j**	9.1^2

2 Use square tables to find the values of these squares.
Set out the working like this.

Example $14.6^2 = (1.46 \times 10)^2 = 2.13 \times 100 = 213$

a	10.4^2	**b**	12.3^2	**c**	13.4^2	**d**	17.3^2	**e**	28.7^2
f	30.6^2	**g**	31.4^2	**h**	32.4^2	**i**	51.4^2	**j**	67.3^2
k	80.5^2	**l**	91.0^2	**m**	98.5^2	**n**	104^2	**o**	127^2
p	253^2	**q**	311^2	**r**	314^2	**s**	324^2	**t**	486^2
u	597^2	**v**	906^2	**w**	1230^2	**x**	2460^2	**y**	2890^2
z	3330^2								

3 Work these *without* using tables.

a	$0.9^2 = \frac{9}{10} \times \frac{9}{10} = \ldots$	**b**	0.8^2	**c**	0.7^2	**d**	0.4^2
e	0.3^2	**f**	0.2^2	**g**	0.1^2		
h	$0.09^2 = \frac{9}{100} \times \frac{9}{100} = \ldots$	**i**	0.07^2	**j**	0.06^2	**k**	0.05^2
l	0.04^2	**m**	0.03^2	**n**	0.01^2		

4 Use square tables to find the values of these squares.
Set out the working like this.

Example $0.623^2 = \left(\frac{6.23}{10}\right)^2 = \frac{38.8}{100} = 0.388$

a	0.725^2	**b**	0.813^2	**c**	0.912^2	**d**	0.405^2
e	0.53^2	**f**	0.63^2	**g**	0.18^2	**h**	0.322^2

Squares

i 0.312^2	j 0.214^2	k 0.125^2	l 0.24^2
m 0.19^2	n 0.101^2	o 0.099^2	p 0.084^2
q 0.0528^2	r 0.0323^2	s 0.0313^2	t 0.025^2
u 0.014^2	v 0.0097^2	w 0.00825^2	x 0.00265^2

5 A mixture

Use square tables to calculate these.

a 7.65^2	b 76.5^2	c 765^2	d 0.765^2
e 4.02^2	f 40.2^2	g 402^2	h 0.402^2
i 2.84^2	j 28.4^2	k 284^2	l 0.284^2
m 12.7^2	n 36.2^2	o 0.445^2	p 0.63^2
q 265^2	r 427^2	s 0.086^2	t 0.055^2
u 0.037^2	v 0.0076^2	w 6460^2	x 1320^2

6 A square has sides of length 18.7 cm. Find the area of the square (correct to 3 significant figures).

7 Another square has sides of length 46.3 cm. Find the area of the square (correct to 3 significant figures).

8 The area A of a circle is given by the formula $A = 3.14 \times r^2$ where r is its radius.
Find the area when the radius is

a 23.6 cm b 47.2 cm c 123 cm d 0.37 cm.

Give the answers correct to three significant figures.

9 The kinetic energy E of an object is given by the formula $E = \frac{1}{2}mv^2$.
Find E (correct to 3 significant figures) when

a $m = 8, v = 18.4$ b $m = 10, v = 42.5$

c $m = 12, v = 0.42$ d $m = 22, v = 0.86$.

10 The distance d travelled by an accelerating car in a time t is given by the formula $d = \frac{1}{2}at^2$.
Find d (correct to 3 significant figures) when

a $a = 6, t = 3.65$ b $a = 14, t = 0.875$

c $a = 7, t = 41.2$ d $a = 5.6, t = 0.55$.

11 The distance d taken by a moving car to come to rest is given by the formula

$$d = \frac{v^2}{2a}.$$

Find d (correct to 3 significant figures) when

a $v = 23.5, a = 5$ b $v = 0.62, a = 5$

c $v = 95.5, a = 4$ d $v = 0.82, a = 6$.

12 A man stands at a height of h metres above sea-level and looks out to sea at the horizon which is x km away. The relation between h and x is given by the formula

$$h = \frac{2x^2}{25}.$$

Find h when

a $x = 35.4$ b $x = 49$ c $x = 124$ d $x = 160$.

Square roots

Part 1 Introduction using factors

Factors of 36

The factors of 36 can be arranged as shown here.

Halfway down the list at $6 \times 6 = 36$, the numbers 'pass each other'.

Therefore 6 is the square root of 36.

It used to be written as $r36$ but now the letter r is distorted into $\sqrt{}$ and it is written $6 = \sqrt{36}$.

$$1 \times 36 = 36$$
$$2 \times 18 = 36$$
$$3 \times 12 = 36$$
$$4 \times 9 = 36$$
$$6 \times 6 = 36$$
$$9 \times 4 = 36$$
$$12 \times 3 = 36$$
$$18 \times 2 = 36$$
$$36 \times 1 = 36$$

Factors of 12

The factors of 12 can be arranged as shown here.

Halfway down the list, the numbers 'pass each other'.

There should be a decimal number n between 3 and 4 such that $n \times n = 12$.

This number n is the square root of 12   i.e. $n = \sqrt{12}$.

$$1 \times 12 = 12$$
$$2 \times 6 = 12$$
$$3 \times 4 = 12$$
$$4 \times 3 = 12$$
$$6 \times 2 = 12$$
$$12 \times 1 = 12$$

To find $\sqrt{12}$ (A calculator would be useful.)

$3 \times 4 = 12$
and $4 \times 3 = 12$.   Take the average of 3 and 4. $\dfrac{3 + 4}{2} = 3.5$

Divide this average into 12 to give $\dfrac{12}{3.5} = 3.4$ to 1 decimal place.

So $3.4 \times 3.5 \simeq 12$
and $3.5 \times 3.4 \simeq 12$.   Take the average of 3.4 and 3.5. $\dfrac{3.4 + 3.5}{2} = 3.45$

Divide this average into 12 to give $\dfrac{12}{3.45} = 3.48$ to 2 decimal places.

Now $3.45 \times 3.48 \simeq 12$
and $3.48 \times 3.45 \simeq 12$.   Take the average of 3.45 and 3.48. $\dfrac{3.45 + 3.48}{2} = 3.465$

Divide this average into 12 to give $\dfrac{12}{3.465} = 3.463$ to 3 decimal places.

So $3.463 \times 3.465 \simeq 12$
and $3.465 \times 3.463 \simeq 12$.   Take the average of these. $\dfrac{3.463 + 3.465}{2} = 3.464$

Divide this average into 12 to give $\dfrac{12}{3.464} = 3.4642$ to 4 decimal places.

Now $3.464 \times 3.4642 \simeq 12$
and $3.4642 \times 3.464 \simeq 12$.   Take the average of these. $\dfrac{3.464 + 3.4642}{2} = 3.4641$

This process can be repeated as many times as required.

At this stage, $\sqrt{12} = 3.4641$.

Square roots

The instructions for finding $\sqrt{A}$ by the above method are laid out in this flow diagram.

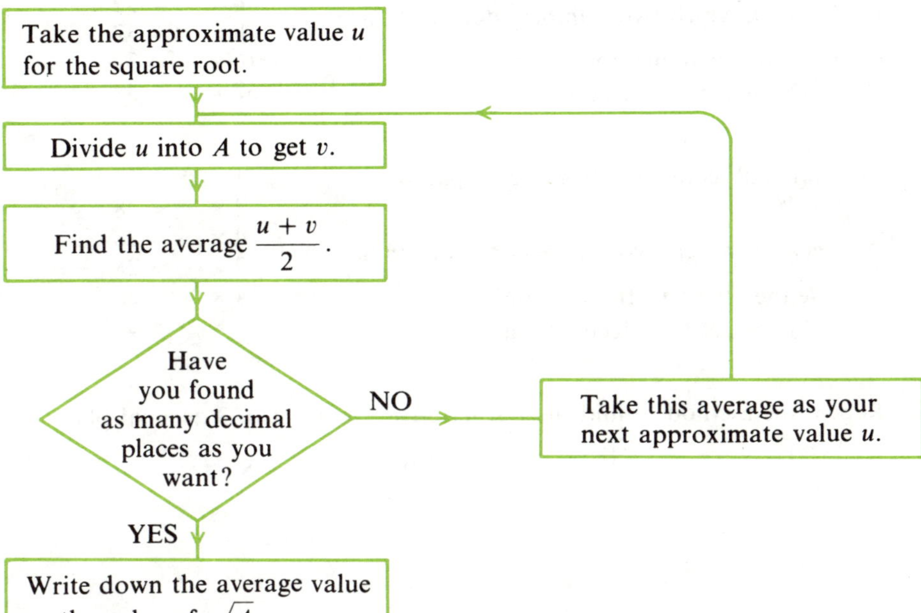

1 Copy and complete these factor pairs for 64.

Write the value of $\sqrt{64}$.

$$1 \times 64 = 64$$
$$2 \times \quad = 64$$
$$4 \times \quad = 64$$
$$8 \times \quad = 64$$
$$16 \times \quad = 64$$
$$32 \times \quad = 64$$
$$64 \times \quad = 64$$

2 Copy and complete these factor pairs for 100.

Write the value of $\sqrt{100}$.

$$1 \times 100 = 100$$
$$2 \times \quad = 100$$
$$4 \times \quad = 100$$
$$5 \times \quad = 100$$
$$10 \times \quad = 100$$
$$20 \times \quad = 100$$
$$25 \times \quad = 100$$
$$50 \times \quad = 100$$
$$100 \times \quad = 100$$

3 Without any working, write the values of

  a $\sqrt{9}$ **b** $\sqrt{25}$ **c** $\sqrt{16}$ **d** $\sqrt{49}$ **e** $\sqrt{81}$ **f** $\sqrt{144}$.

4 a Copy and complete these factor pairs for 40.

  b Between which two numbers does $\sqrt{40}$ lie?

  c Use the above method to find $\sqrt{40}$ correct to 2 decimal places.

$$1 \times 40 = 40$$
$$2 \times \quad = 40$$
$$4 \times \quad = 40$$
$$5 \times \quad = 40$$
$$8 \times \quad = 40$$
$$10 \times \quad = 40$$
$$20 \times \quad = 40$$
$$40 \times \quad = 40$$

Square roots

5 a Copy and complete these factor pairs
      for 32.

   b Between which two numbers does $\sqrt{32}$ lie?

   c Use the above method to find
      $\sqrt{32}$ correct to 2 decimal places.

$$1 \times 32 = 32$$
$$2 \times \quad = 32$$
$$4 \times \quad = 32$$
$$8 \times \quad = 32$$
$$16 \times \quad = 32$$
$$32 \times \quad = 32$$

6 a Copy and complete these factor pairs
      for 28.

   b Between which two numbers does $\sqrt{28}$ lie?

   c Use the above method to find
      $\sqrt{28}$ correct to 2 decimal places.

$$1 \times 28 = 28$$
$$2 \times \quad = 28$$
$$4 \times \quad = 28$$
$$7 \times \quad = 28$$
$$14 \times \quad = 28$$
$$28 \times \quad = 28$$

7 Repeat the method to find these square roots correct to 3 decimal places.

   a $\sqrt{30}$ b $\sqrt{18}$ c $\sqrt{50}$ d $\sqrt{35}$ e $\sqrt{8}$

Part 2 Introduction using graphs

Factors of 36

Let each factor pair give a
point (x, y) which is plotted
as shown. The points are
then joined to form a
smooth curve.

x	y	point
1×36	$\rightarrow$	$(1, 36)$
2×18	$\rightarrow$	$(2, 18)$
3×12	$\rightarrow$	$(3, 12)$
4×9	$\rightarrow$	$(4, 9)$
9×4	$\rightarrow$	$(9, 4)$
12×3	$\rightarrow$	$(12, 3)$
18×2	$\rightarrow$	$(18, 2)$
36×1	$\rightarrow$	$(36, 1)$

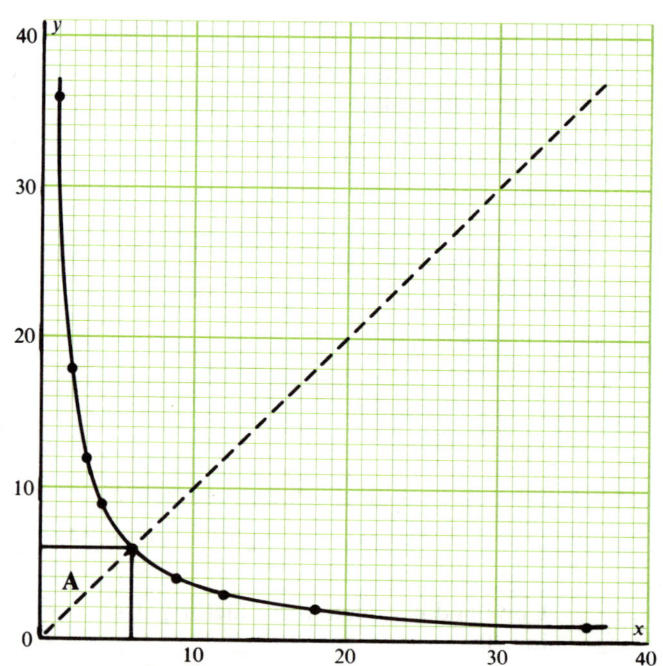

$\sqrt{36}$ is found at the point (x, y), where the dotted line $y = x$ cuts the curve.

By looking at the graph drawn here, write
   a the value of $\sqrt{36}$ b the area of square **A**.

Square roots

Factors of 12

Let all the factor pairs give points (x, y) which can be plotted and joined together to form a smooth curve.

x	y	point
1×12	$\rightarrow$	$(1, 12)$
2×6	$\rightarrow$	$(2, 6)$
3×4	$\rightarrow$	$(3, 4)$
4×3	$\rightarrow$	$(4, 3)$
6×2	$\rightarrow$	$(6, 2)$
12×1	$\rightarrow$	$(12, 1)$

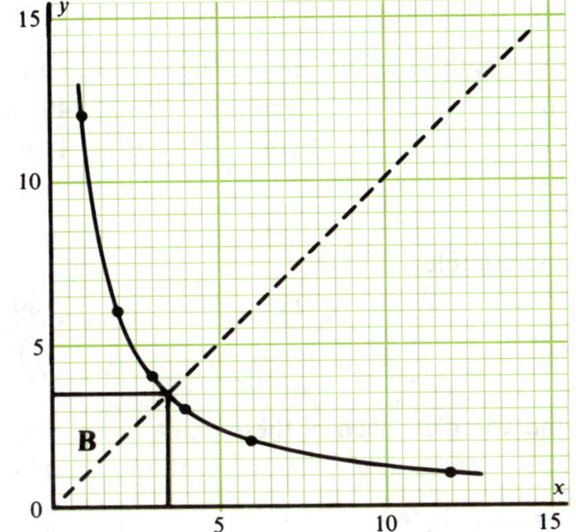

$\sqrt{12}$ is found at the point of intersection of the dotted line with the curve.

Look at the graph drawn here, and write
  a   the value of $\sqrt{12}$ b   the area of square **B**.

1   Copy and complete these factor pairs for 18.

Plot a curve and find the value of $\sqrt{18}$ as accurately as possible.

$$x \times y = 18$$
$$1 \times 18 = 18$$
$$2 \times 9 = 18$$
$$3 \times 6 =$$
$$6 \times 3 =$$
$$9 \times 2 =$$
$$18 \times 1 =$$

2   Copy and complete these factor pairs for 10.

Plot a curve and find the value of $\sqrt{10}$ as accurately as possible.

$$x \times y = 10$$
$$1 \times 10 = 10$$
$$2 \times 5 =$$
$$5 \times 2 =$$
$$10 \times 1 =$$

3   Use these pairs of factors to find $\sqrt{15}$ and $\sqrt{20}$ from a graph.

  a
$$1 \times 15 = 15$$
$$3 \times 5 = 15$$
$$5 \times 3 = 15$$
$$15 \times 1 = 15$$

  b
$$1 \times 20 = 20$$
$$2 \times 10 = 20$$
$$4 \times 5 = 20$$
$$5 \times 4 = 20$$
$$10 \times 2 = 20$$
$$20 \times 1 = 20$$

4   First write all the factor pairs then use a graphical method to find

  a $\sqrt{24}$ b $\sqrt{28}$ c $\sqrt{8}$.

5   Why is this graphical method no use in finding

  a $\sqrt{7}$ b $\sqrt{11}$ c $\sqrt{13}$?

Square roots

Part 3

1 Write the length of a side of a square, if it has an area of

   a   9 cm²   b   25 cm²   c   49 cm²

   d   100 cm²   e   81 cm²   f   1 cm²

   g $\frac{1}{4}$ cm².

2 Copy and complete.

   a $\sqrt{9} = \ldots$ b $\sqrt{25} = \ldots$ c $\sqrt{49} = \ldots$ d $\sqrt{100} = \ldots$

   e $\sqrt{81} = \ldots$ f $\sqrt{1} = \ldots$ g $\sqrt{\frac{1}{4}} = \ldots$

3 Copy and complete each of these pairs.

   a $6^2 = \ldots, \quad \sqrt{36} = \ldots$ b $4^2 = \ldots, \quad \sqrt{16} = \ldots$

   c $8^2 = \ldots, \quad \sqrt{64} = \ldots$ d $2^2 = \ldots, \quad \sqrt{4} = \ldots$

   e $10^2 = \ldots, \quad \sqrt{100} = \ldots$ f $12^2 = \ldots, \quad \sqrt{144} = \ldots$

   g $11^2 = \ldots, \quad \sqrt{121} = \ldots$ h $(\frac{1}{3})^2 = \ldots, \quad \sqrt{\frac{1}{9}} = \ldots$

4 Break each of these numbers into its prime factors, and hence find the square root. (Check the answers by long multiplication).

   Example $1936 = 2 \times 2 \times 2 \times 2 \times 11 \times 11$

$$\sqrt{1936} = 2 \times 2 \times 11 = 44$$

a	1764	b	900	c	196	d	324
e	1296	f	1600	g	225	h	2025
i	5625	j	576	k	7056	l	2304
m	3969	n	2401	o	17 424	p	6084
q	4225	r	9801	s	33 124		

5 Write the square roots of these fractions.

   a $\sqrt{\frac{1}{9}}$ b $\sqrt{\frac{1}{16}}$ c $\sqrt{\frac{1}{25}}$ d $\sqrt{\frac{1}{49}}$

   e $\sqrt{\frac{4}{9}}$ f $\sqrt{\frac{9}{25}}$ g $\sqrt{\frac{16}{81}}$ h $\sqrt{\frac{49}{100}}$

   i $\sqrt{2\frac{1}{4}}$ j $\sqrt{6\frac{1}{4}}$ k $\sqrt{1\frac{7}{9}}$ l $\sqrt{3\frac{1}{16}}$

   m $\sqrt{3\frac{6}{25}}$ n $\sqrt{2\frac{7}{9}}$ o $\sqrt{12\frac{1}{4}}$ p $\sqrt{5\frac{19}{25}}$

   q $\sqrt{7\frac{9}{16}}$ r $\sqrt{\frac{45}{80}}$ s $\sqrt{2\frac{3}{12}}$ t $\sqrt{1\frac{14}{18}}$

6 Write these decimals as fractions and find their square roots.

   Example $\sqrt{0.36} = \sqrt{\frac{36}{100}} = \frac{6}{10} = 0.6$

   a $\sqrt{0.16}$ b $\sqrt{0.25}$ c $\sqrt{0.81}$ d $\sqrt{0.04}$

   e $\sqrt{0.09}$ f $\sqrt{0.01}$ g $\sqrt{1.44}$ h $\sqrt{1.21}$

   i $\sqrt{0.0064}$ j $\sqrt{0.0036}$ k $\sqrt{0.0049}$ l $\sqrt{0.0009}$

   m $\sqrt{0.0144}$ n $\sqrt{0.0121}$ o $\sqrt{0.0169}$ p $\sqrt{0.0225}$

Square roots

Part 4 The graph of $y = \sqrt{x}$

1

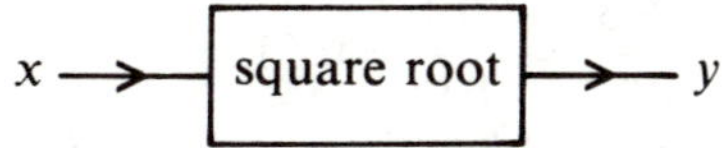

Copy and complete this table for the equation $y = \sqrt{x}$.

x	0	1	4	9	16	25	36	49	64	81	100
y											

Plot the graph of $y = \sqrt{x}$ onto squared paper.

Use this graph to write the values of these square roots as accurately as possible.

a $\sqrt{90}$ b $\sqrt{70}$ c $\sqrt{55}$ d $\sqrt{45}$ e $\sqrt{87}$

f $\sqrt{60}$ g $\sqrt{35}$ h $\sqrt{30}$ i $\sqrt{27}$ j $\sqrt{20}$

k $\sqrt{18}$ l $\sqrt{12}$ m $\sqrt{10}$ n $\sqrt{8}$ o $\sqrt{5}$

2 Use the graph to find approximate values of

a $\sqrt{(\sqrt{95})}$ b $\sqrt{(\sqrt{86})}$ c $\sqrt{(\sqrt{72})}$.

Part 5 Using tables

1 Use **square tables** forwards and backwards to calculate these.

a 2.07^2, $\sqrt{4.28}$ b 4.13^2, $\sqrt{17.1}$ c 7.08^2, $\sqrt{50.1}$

d 9.25^2, $\sqrt{85.6}$ e 11.0^2, $\sqrt{121}$ f 12.5^2, $\sqrt{156}$

g 14.7^2, $\sqrt{216}$ h 20.0^2, $\sqrt{400}$ i 20.8^2, $\sqrt{433}$

j $\sqrt{16.5}$ k $\sqrt{25.2}$ l $\sqrt{87.0}$ m $\sqrt{93.1}$

n $\sqrt{3.69}$ o $\sqrt{15.8}$ p $\sqrt{49.3}$ q $\sqrt{99.6}$

2 Use **square-root tables** to find the values of these.

a $\sqrt{6.55}$ b $\sqrt{65.5}$ c $\sqrt{4.82}$ d $\sqrt{48.2}$

e $\sqrt{9.06}$ f $\sqrt{90.6}$ g $\sqrt{1.73}$ h $\sqrt{17.3}$

i $\sqrt{50.2}$ j $\sqrt{2.7}$ k $\sqrt{8.4}$ l $\sqrt{4.1}$

m $\sqrt{67}$ n $\sqrt{82}$ o $\sqrt{34}$ p $\sqrt{49}$

3 Numbers greater than 100

Example $\sqrt{418} = \sqrt{4.18 \times 100} = 2.04 \times 10 = 20.4$

Calculate, using square-root tables.

a $\sqrt{524}$ b $\sqrt{847}$ c $\sqrt{639}$ d $\sqrt{142}$

e $\sqrt{350}$ f $\sqrt{130}$ g $\sqrt{600}$ h $\sqrt{800}$

i $\sqrt{1320}$ j $\sqrt{2750}$ k $\sqrt{3840}$ l $\sqrt{7910}$

m $\sqrt{8040}$ n $\sqrt{6010}$ o $\sqrt{6500}$ p $\sqrt{4200}$

q $\sqrt{1600}$ r $\sqrt{7000}$ s $\sqrt{2000}$ t $\sqrt{62\,500}$

u $\sqrt{85\,600}$ v $\sqrt{94\,400}$ w $\sqrt{12\,500}$ x $\sqrt{33\,700}$

y $\sqrt{176\,000}$ z $\sqrt{332\,000}$

Square roots

4   Numbers less than 1

Example $\sqrt{0.263} = \sqrt{\dfrac{26.3}{100}} = \dfrac{5.13}{10} = 0.513$

Calculate, using square root tables.

a $\sqrt{0.375}$	b $\sqrt{0.632}$	c $\sqrt{0.155}$	d $\sqrt{0.203}$
e $\sqrt{0.702}$	f $\sqrt{0.73}$	g $\sqrt{0.69}$	h $\sqrt{0.52}$
i $\sqrt{0.4}$	j $\sqrt{0.9}$	k $\sqrt{0.7}$	l $\sqrt{0.824}$
m $\sqrt{0.063}$	n $\sqrt{0.052}$	o $\sqrt{0.036}$	p $\sqrt{0.081}$
q $\sqrt{0.0773}$	r $\sqrt{0.0614}$	s $\sqrt{0.06}$	t $\sqrt{0.09}$
u $\sqrt{0.05}$	v $\sqrt{0.00375}$	w $\sqrt{0.00513}$	x $\sqrt{0.0064}$
y $\sqrt{0.0095}$	z $\sqrt{0.00072}$		

5   Find the values of these.

a $\sqrt{0.16}$	b $\sqrt{0.016}$	c $\sqrt{0.36}$	d $\sqrt{0.036}$
e $\sqrt{0.81}$	f $\sqrt{0.081}$	g $\sqrt{0.4}$	h $\sqrt{0.04}$
i $\sqrt{0.9}$	j $\sqrt{0.09}$	k $\sqrt{0.08}$	l $\sqrt{0.0843}$
m $\sqrt{0.0724}$	n $\sqrt{0.0631}$		

6   Revision

Find the values of these.

a $\sqrt{420}$	b $\sqrt{4200}$	c $\sqrt{0.42}$	d $\sqrt{0.042}$
e $\sqrt{874}$	f $\sqrt{8740}$	g $\sqrt{0.874}$	h $\sqrt{0.0874}$
i $\sqrt{469}$	j $\sqrt{5370}$	k $\sqrt{0.63}$	l $\sqrt{0.07}$
m $\sqrt{740}$	n $\sqrt{0.457}$	o $\sqrt{0.0632}$	p $\sqrt{0.0707}$

7   If a square has an area of 46.7 cm², what is its length of side?

8   Another square has an area of 467 cm². Find its length of side.

9   The radius r of a circle is given by the formula

$$r = \sqrt{\dfrac{A}{\pi}}.$$

Find r when   a $A = 12.9, \pi = 3$ b $A = 645, \pi = 3$.

10   A formula to calculate the electric current I flowing through a resistance R is

$$I = \sqrt{\dfrac{W}{R}}.$$

Find I when   a $W = 175, R = 2$ b $W = 1240, R = 5$.

11   The velocity V of a mass m which has an energy E is given by the formula

$$V = \sqrt{\dfrac{2E}{m}}.$$

Find V when   a $E = 436, m = 10$ b $E = 2560, m = 4$.

Square roots

12 The time t taken for a car to accelerate a distance d from rest is given by the formula

$$t = \sqrt{\frac{2d}{a}}.$$

Find t when   **a** $d = 0.385, a = 5$ **b** $d = 0.07, a = 4$.

13 The radius r of a sphere with a surface area A is given by the formula

$$r = \sqrt{\frac{A}{4\pi}}.$$

Find r when   **a** $\pi = 3, A = 1500$ **b** $\pi = 3, A = 0.45$.

14 The radius r of a cylinder is given by the formula

$$r = \sqrt{\frac{V}{\pi h}}.$$

Find r when   **a** $V = 864, \pi = 3, h = 2$ **b** $V = 0.738, \pi = 3, h = 3$.

15 If you stand at a height of h metres above sea-level, the horizon is x kilometres away, where $x = \dfrac{7\sqrt{h}}{2}$.

Find x when   **a** $h = 8.5$ **b** $h = 623$ **c** $h = 2120$.

The laws of indices

Part 1 Introduction

1 Write the area of a square which has sides of
   a 4 cm   b 7 cm.

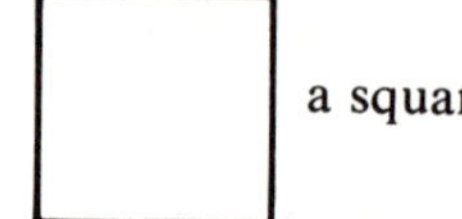
a square

2 Write the length of side of a square which has an area of
   a 25 cm^2 b 81 cm^2.

3 Write the volume of a cube which has edges of
   a 3 cm   b 5 cm.

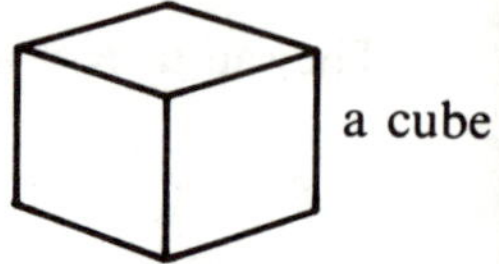
a cube

4 Write the length of edge of a cube which has a volume of
   a 8 cm^3 b 1000 cm^3.

5 Copy and complete these pairs.

   a $6^2 = \ldots$, $\sqrt{36} = \ldots$ b $4^2 = \ldots$, $\sqrt{16} = \ldots$

   c $2^3 = \ldots$, $\sqrt[3]{8} = \ldots$ d $5^3 = \ldots$, $\sqrt[3]{125} = \ldots$

   e $2^4 = \ldots$, $\sqrt[4]{16} = \ldots$ f $10^4 = \ldots$, $\sqrt[4]{10\,000} = \ldots$

6 Write the value of each of these.

   a $\sqrt{25}$ b $\sqrt{100}$ c $\sqrt[3]{8}$ d $\sqrt[3]{27}$ e $\sqrt[3]{1000}$

   f $\sqrt[4]{16}$ g $\sqrt[4]{81}$ h $\sqrt[5]{32}$ i $\sqrt[6]{64}$ j $\sqrt{64}$

   k $\sqrt[3]{64}$ l $\sqrt{400}$ m $\sqrt{\frac{1}{4}}$ n $\sqrt[3]{\frac{1}{8}}$ o $\sqrt{\frac{1}{25}}$

   p $\sqrt[3]{\frac{1}{125}}$ q $\sqrt{\frac{4}{9}}$ r $\sqrt[3]{\frac{8}{27}}$ s $\sqrt[5]{\frac{1}{32}}$ t $\sqrt[4]{\frac{16}{81}}$

   u $\sqrt{\frac{9}{100}}$ v $\sqrt[3]{\frac{27}{1000}}$ w $\sqrt{2\frac{1}{4}}$ x $\sqrt[3]{3\frac{3}{8}}$ y $\sqrt[3]{2\frac{10}{27}}$

   z $\sqrt[4]{5\frac{1}{16}}$

7 Calculate the value of each of these.

   a 6^3 b 7^3 c 5^4 d 10^5

   e $\sqrt[3]{216}$ f $\sqrt[3]{343}$ g $\sqrt[4]{625}$ h $\sqrt[5]{100\,000}$

   i 25^2 j 40^2 k 8^3 l 9^4

   m $\sqrt{625}$ n $\sqrt{1600}$ o $\sqrt[3]{512}$ p $\sqrt[4]{6561}$

8

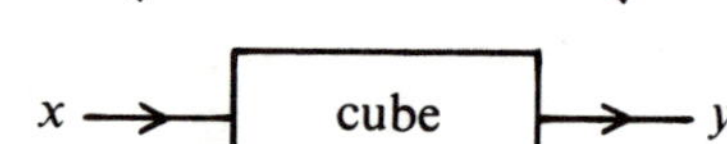

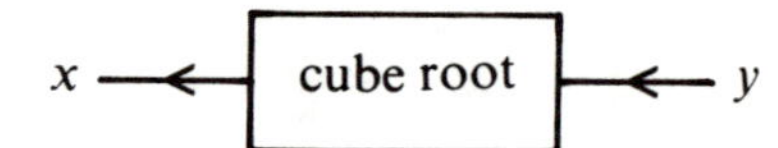

Copy and complete this table for the equation $y = x^3$.

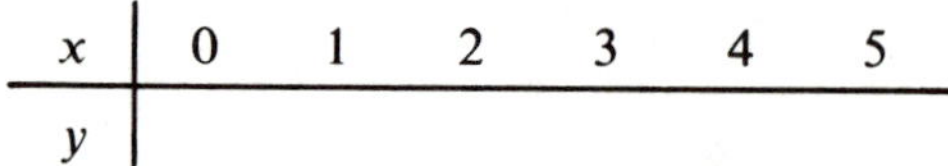

x	0	1	2	3	4	5
y						

Plot the graph of $y = x^3$ onto squared paper.

Use this graph to write the value of each of these as accurately as possible.

   a 1.5^3 b 2.5^3 c 2.8^3 d 3.1^3 e 4.3^3

   f 4.8^3 g 0.5^3 h 0.4^3 i $\sqrt[3]{100}$ j $\sqrt[3]{80}$

   k $\sqrt[3]{75}$ l $\sqrt[3]{45}$ m $\sqrt[3]{38}$ n $\sqrt[3]{62}$ o $\sqrt[3]{20}$

   p 1.35^3 q 2.65^3 r 4.74^3

The laws of indices

9 These are the first seven prime numbers 2, 3, 5, 7, 11, 13, 17.
Break each of these numbers into prime factors and hence find the cube root of each number.

Example $\quad 21\,952 = 2 \times 2 \times 2 \times 2 \times 2 \times 2 \times 2 \times 7 \times 7 \times 7$

$$\sqrt[3]{21\,952} = 2 \times 2 \times 7 = 28$$

a	216	b	512	c	3375	d	1728	e	91 125
f	2744	g	74 088	h	19 683	i	9261	j	42 875

10 Use the same method to find the values of these.

    a $\sqrt[4]{1296}$ b $\sqrt[4]{50\,625}$ c $\sqrt[5]{248\,832}$ d $\sqrt[5]{537\,824}$

Part 2 The laws of indices

Law I $\quad x^a \times x^b = x^{a+b}$

Simplify (but do not calculate) the values of these.

1	$2^7 \times 2^3$	5	$5^6 \times 5^4 \times 5^3$	9	$4 \times 4 \times 4^3$
2	$3^5 \times 3^4$	6	$7^3 \times 7^4 \times 7^2$	10	$6^3 \times 6^5 \times 6^{-4}$
3	$6^8 \times 6^3$	7	$2^3 \times 2^7 \times 2$	11	$2^9 \times 2^3 \times 2^{-6}$
4	$8^3 \times 8^5$	8	$3^5 \times 3 \times 3^2$	12	$5^3 \times 5^7 \times 5^{-4}$

Simplify these.

13	$a^3 \times a^4$	16	$x^6 \times x$	19	$e^2 \times e^5 \times e^{-3}$
14	$m^6 \times m^7$	17	$y^4 \times y^3 \times y$	20	$z^5 \times z^{-2} \times z$
15	$n^3 \times n^5 \times n^2$	18	$s \times s \times s^4$	21	$a^4 \times a \times a^{-1}$

Find the value of n in each of these.

22	$2^4 \times 2^5 = 2^n$	24	$6^7 \times 6^n = 6^{11}$	26	$5^8 \times 5^n = 5^7$
23	$3^6 \times 3^n = 3^8$	25	$4^4 \times 4^n = 4^5$	27	$8^n \times 8^n = 8^6$

Law II $\quad x^a \div x^b = x^{a-b}$

Simplify (but do not calculate) the values of these.

1	$\dfrac{2^9}{2^4}$	6	$\dfrac{12^4}{12}$	11	$\dfrac{7^8}{7^2 \times 7^3}$
2	$\dfrac{8^6}{8^2}$	7	$\dfrac{4^7 \times 4^5}{4^6}$	12	$\dfrac{4^{10}}{4 \times 4^5}$
3	$\dfrac{5^7}{5^4}$	8	$\dfrac{5^8 \times 5^4}{5^7}$	13	$\dfrac{3^6 \times 3^5}{3^7 \times 3^2}$
4	$\dfrac{3^9}{3^8}$	9	$\dfrac{6^5 \times 6^7 \times 6^3}{6^{12}}$	14	$\dfrac{10^2 \times 10^8}{10^6 \times 10^3}$
5	$\dfrac{4^6}{4}$	10	$\dfrac{2 \times 2^8 \times 2^4}{2^{10}}$	15	$\dfrac{2^7 \times 2^9 \times 2^3}{2^8 \times 2^{10}}$

The laws of indices

Simplify these.

16 $\dfrac{a^8}{a^6}$

17 $\dfrac{c^7}{c^4}$

18 $\dfrac{d^5}{d^3}$

19 $\dfrac{x^6}{x}$

20 $\dfrac{y^8}{y}$

21 $\dfrac{m^2 \times m^4}{m^3}$

22 $\dfrac{a^8 \times a^7}{a^5}$

23 $\dfrac{z \times z^6}{z^4}$

24 $\dfrac{k^7 \times k}{k^6}$

25 $\dfrac{n^9}{n^4 \times n^3}$

26 $\dfrac{s^{12}}{s^5 \times s^4}$

27 $\dfrac{a^6 \times a^9}{a^4 \times a^8}$

Find the value of n in each of these.

28 $\dfrac{8^6}{8^n} = 8^4$

29 $\dfrac{7^n}{7^6} = 7^3$

30 $\dfrac{a^n}{a} = a^5$

31 $\dfrac{x^4 \times x^5}{x^n} = x^7$

32 $\dfrac{m^6 \times m^n}{m^4} = m^5$

33 $\dfrac{t^3 \times t^n}{t^5} = t^4$

Law III $(x^a)^b = x^{ab}$

1 Copy and complete these using **Law I**.
   a $(3^5)^4 = 3^5 \times 3^5 \times 3^5 \times 3^5 = \ldots$
   b $(2^3)^5 = 2^3 \times 2^3 \times 2^3 \times 2^3 \times 2^3 = \ldots$

Simplify (but do not calculate) the values of these.

2 $(2^6)^5$

3 $(4^3)^7$

4 $(3^5)^2$

5 $(5^2)^4$

6 $(12^3)^5$

7 $(7^2)^3$

8 $(2^7)^2 \times 2^4$

9 $(6^3)^2 \times 6^5$

10 $3^3 \times (3^2)^4$

11 $4^5 \times (4^3)^4$

12 $\dfrac{(5^3)^2}{5^4}$

13 $\dfrac{(10^5)^3}{10^9}$

14 $\dfrac{(8^6)^4}{8^{10}}$

15 $\dfrac{7^9}{(7^2)^3}$

16 $\dfrac{6^4 \times (6^2)^3}{6^5}$

Simplify these.

17 $(u^3)^4$

18 $(x^4)^2$

19 $(y^6)^3$

20 $z^4 \times (z^3)^2$

21 $m^6 \times (m^2)^2$

22 $\dfrac{(n^4)^2}{n^5}$

23 $\dfrac{(a^4)^3}{a}$

24 $\dfrac{s^8}{(s^2)^2}$

25 $\dfrac{(r^3)^3 \times r^6}{r^9}$

The laws of indices

Find the value of n in these.

26 $(4^2)^n = 4^8$ **28** $(2^n)^5 = 2^{15}$ **30** $\dfrac{(x^3)^5}{x^n} = x^2$ **32** $\dfrac{(u^n)^2}{u^4} = u^6$

27 $(3^3)^n = 3^{12}$ **29** $\dfrac{(a^6)^3}{a^n} = a^8$ **31** $\dfrac{t^n}{(t^3)^2} = t^2$

Law IV $\quad x^0 = 1$

1 **a** Copy this statement and use **Law I** to complete it. $\quad 3^7 \times 3^0 = \ldots$

b What number will fill the gap in this statement? $\quad 3^7 \times \ldots = 3^7$

c Hence write the value of 3^0.

Simplify these.

2 4^0 **5** $(\frac{1}{2})^0$ **8** $3^6 \times 3^{-6}$ **11** $\dfrac{2^8 \times 2^5}{2^9 \times 2^4}$

3 9^0 **6** $\dfrac{7^8}{7^8}$ **9** $8^3 \times 8^{-3}$ **12** $\dfrac{4 \times 4^3}{4^4}$

4 $(2.6)^0$ **7** $\dfrac{4^3}{4^3}$ **10** $\dfrac{5^6 \times 5^3}{7^5 \times 7^4}$

Law V $\quad x^{-a} = \dfrac{1}{x^a}$

1 **a** Copy this statement and use **Law I** to complete it. $\quad 3^7 \times 3^{-5} = \ldots$

b Copy this statement and use **Law II** to complete it. $\quad \dfrac{3^7}{3^5} = \ldots$

c What can you now deduce about 3^{-5} and $\dfrac{1}{3^5}$?

Simplify these and calculate their values.

Example $\quad 9^{-2} = \dfrac{1}{9^2} = \dfrac{1}{81}$

2 6^{-2} **10** $\dfrac{6^5}{6^7}$ **18** 4×10^{-3}

3 7^{-2} **11** $\dfrac{2^5}{2^8}$ **19** 8×10^{-2}

4 2^{-3} **12** $\dfrac{3^7}{3^9}$ **20** 0.7×10^{-2}

5 3^{-3} **13** 160×10^{-2} **21** 0.35×10^{-3}

6 3^{-4} **14** 137.4×10^{-3} **22** 0.4×10^{-5}

7 10^{-2} **15** 262.5×10^{-2} **23** 18×10^{-3}

8 2^{-5} **16** 36.9×10^{-1} **24** 850×10^{-1}

9 1^{-4} **17** 12×10^{-2}

The laws of indices

Law VI $x^{\frac{1}{a}} = \sqrt[a]{x}$

1 **a** Copy this statement and use **Law I** to complete it. $9^{\frac{1}{2}} \times 9^{\frac{1}{2}} = \ldots$

   b Work out the square roots to complete this statement. $\sqrt{9} \times \sqrt{9} = \ldots$

   c What can you now say about $9^{\frac{1}{2}}$ and $\sqrt{9}$?

2 **a** Copy this statement and use **Law I** to complete it. $8^{\frac{1}{3}} \times 8^{\frac{1}{3}} \times 8^{\frac{1}{3}} = \ldots$

   b Work out the cube roots to complete this statement. $\sqrt[3]{8} \times \sqrt[3]{8} \times \sqrt[3]{8} = \ldots$

   c What can you now say about $8^{\frac{1}{3}}$ and $\sqrt[3]{8}$?

Write the values of these.

3 $25^{\frac{1}{2}}$	**7** $1000^{\frac{1}{3}}$	**11** $81^{\frac{1}{2}}$	**15** $32^{\frac{1}{5}}$
4 $36^{\frac{1}{2}}$	**8** $64^{\frac{1}{2}}$	**12** $144^{\frac{1}{2}}$	**16** $64^{\frac{1}{6}}$
5 $100^{\frac{1}{2}}$	**9** $64^{\frac{1}{3}}$	**13** $1^{\frac{1}{2}}$	**17** $125^{\frac{1}{3}}$
6 $27^{\frac{1}{3}}$	**10** $16^{\frac{1}{2}}$	**14** $16^{\frac{1}{4}}$	**18** $81^{\frac{1}{4}}$

Simplify these.

19 $(a^4)^{\frac{1}{2}}$	**23** $\sqrt[3]{y^6}$	**27** $\sqrt{z^8}$	**31** $\sqrt[3]{27y^{12}}$
20 $(c^8)^{\frac{1}{2}}$	**24** $\sqrt[3]{z^{15}}$	**28** $\sqrt{a^{12}}$	**32** $\sqrt[4]{16z^8}$
21 $(x^6)^{\frac{1}{3}}$	**25** $\sqrt[4]{m^8}$	**29** $\sqrt{9x^6}$	
22 $(y^{10})^{\frac{1}{5}}$	**26** $\sqrt[4]{n^{20}}$	**30** $\sqrt[3]{8x^6}$	

A mixture

Find the values of these.

1 $\dfrac{3^8}{3^6}$	**10** $(3^2)^2$	**19** $25^{\frac{1}{2}} \times 49^{\frac{1}{2}}$	**28** 0.65×10^2
2 $\dfrac{2^7}{2^4}$	**11** $\dfrac{(8^3)^4}{8^{10}}$	**20** $\dfrac{81^{\frac{1}{2}}}{27^{\frac{1}{3}}}$	**29** 0.04×10^3
3 $2^2 \times 2^3$	**12** $\dfrac{(7^4)^2}{7^7}$	**21** 4^{-2}	**30** 26.5×10^{-2}
4 3×3^2	**13** $\dfrac{5^2 \times (5^3)^2}{5^8}$	**22** 5^{-3}	**31** 8.5×10^{-3}
5 $\dfrac{4^6 \times 4^3}{4^7}$	**14** $\dfrac{3^4 \times 3^7}{3^{11}}$	**23** 2^{-5}	**32** 0.7×10^{-4}
6 $\dfrac{5^8 \times 5}{5^6}$	**15** $\dfrac{9^4}{9^6 \times 9^{-2}}$	**24** $\dfrac{6^2}{6^4}$	**33** 0.7×10^4
7 $\dfrac{6^7}{6^4 \times 6^2}$	**16** $16^{\frac{1}{2}}$	**25** $\dfrac{7^3}{7^4}$	**34** 1.6×10^{-5}
8 $\dfrac{8^9}{8^6 \times 8}$	**17** $16^{\frac{1}{4}}$	**26** 8.42×10^2	**35** 1.6×10^5
9 $(2^2)^3$	**18** $8^{\frac{1}{3}} \times 9^{\frac{1}{2}}$	**27** 7.9×10^3	**36** 0.008×10^4

The laws of indices

37 a $(2^2)^2$ b $(3^2)^2$ c $(2^4)^2$ d $(10^2)^3$ e $(1^3)^4$

f $(4^3)^2$ g $(5^3)^2$ h $(2^3)^2$

38 a 4^{-2} b 2^{-4} c 3^{-2} d 2^{-3} e 10^{-3}

f 2^{-6} g 5^{-2} h 4^{-4}

39 a $4^{\frac{1}{2}}$ b $9^{\frac{1}{2}}$ c $49^{\frac{1}{2}}$ d $8^{\frac{1}{3}}$ e $1000^{\frac{1}{3}}$

f $125^{\frac{1}{3}}$ g $16^{\frac{1}{4}}$ h $81^{\frac{1}{4}}$ i $32^{\frac{1}{5}}$ j $64^{\frac{1}{6}}$

40 a $9^{\frac{3}{2}}$ b $16^{\frac{3}{2}}$ c $100^{\frac{3}{2}}$ d $8^{\frac{2}{3}}$ e $27^{\frac{2}{3}}$

f $1000^{\frac{2}{3}}$ g $32^{\frac{3}{5}}$ h $64^{\frac{2}{3}}$ i $25^{\frac{3}{2}}$ j $81^{\frac{3}{4}}$

k $32^{\frac{2}{5}}$ l $64^{\frac{5}{6}}$ m $36^{\frac{3}{2}}$ n $125^{\frac{2}{3}}$ o $64^{\frac{7}{6}}$

p $16^{\frac{5}{4}}$

41 a $8^{-\frac{4}{3}}$ b $25^{-\frac{3}{2}}$ c $16^{-\frac{3}{4}}$ d $125^{-\frac{2}{3}}$ e $4^{-\frac{5}{2}}$

f $81^{-\frac{3}{2}}$ g $100^{-\frac{3}{2}}$ h $27^{-\frac{2}{3}}$ i $32^{-\frac{4}{5}}$ j $1000^{-\frac{2}{3}}$

k $9^{-\frac{3}{2}}$ l $81^{-\frac{3}{4}}$ m $8^{-\frac{5}{3}}$ n $625^{-\frac{3}{4}}$ o $144^{-\frac{3}{2}}$

42 a 4×4^3 b $\dfrac{8^7 \times 8^4}{8^9}$ c $\dfrac{5^7}{5^2 \times 5^3}$ d $\dfrac{2^7 \times 2}{2^8}$

e $4^7 \times 4^{-5}$ f $3^4 \times 3^{-3}$ g $\dfrac{2^8}{2^9}$ h 2^{-1}

i $(2^{-1})^2$ j $16^{\frac{1}{2}}$ k $16^{-\frac{1}{2}}$ l $\left(16^{\frac{1}{2}}\right)^3$

m $16^{\frac{3}{2}}$ n $16^{-\frac{3}{2}}$ o $\sqrt[3]{8^2}$ p $8^{\frac{2}{3}}$

q $8^{-\frac{2}{3}}$ r $1^{\frac{4}{3}}$ s $(0.1)^{-1}$ t $\dfrac{(4^2)^3 \times 4^{-4}}{4^2}$

Part 3 Standard form

1 Write these numbers in full without using indices.

a 2.6×10^4 b 4.72×10^5 c 6.72×10^3

d 8.76×10^{-3} e 2.04×10^{-6} f 1.2×10^{-2}

g 8.9×10^5 h 1.02×10^{-4} i 9.8×10^5

j 2.3×10^{-3} k 4.25×10^{-2} l 3.05×10^6

m 1.2×10^5 n 3×10^4 o 2×10^6

p 2×10^{-6} q 1×10^{-3} r 9×10^{-4}

2 Write these numbers in standard form.

a 416 000   b 2 400 000   c 370 000 000

d 0.00025   e 0.00463   f 0.000068

g 1 000 000   h 0.000001   i 500 000

j 450 000   k 292 000   l 0.0075

m 0.000025   n 0.017   o 1700

p 6 850 000   q 0.00000685   r 0.004

The laws of indices

3 Write these items of information in standard form.

  a The distance from the Earth to the Sun is 93 000 000 miles.

  b The diameter of the sun is 865 000 miles.

  c The length of the Earth's equator is approximately 40 000 000 metres.

  d The diameter of a molecule of hydrogen is 0.00000024 mm.

  e The wavelength of sodium light is 0.00000059 metres.

  f The population of India is 626 000 000.

  g The population of the United Kingdom is 56 000 000.

  h The electric charge on an electron is 0.00000000000016 μC.

  i The number of molecules in 1 mole of any material is
  602 300 000 000 000 000 000 000.

  j The mass of a neutron is 0.000000000000000000001675 milligrams.

4 Find the value of n in these.

  a $75\,000 = 7.5 \times 10^n$

  b $9200 = 9.2 \times 10^n$

  c $160\,000 = 1.6 \times 10^n$

  d $0.00084 = 8.4 \times 10^n$

  e $0.0062 = 6.2 \times 10^n$

  f $0.00007 = 7 \times 10^n$

  g $0.098 = 9.8 \times 10^n$

  h $980 = 9.8 \times 10^n$

  i $76\,000\,000 = 7.6 \times 10^n$

  j $1\,000\,000 = 1 \times 10^n$

  k $0.000001 = 1 \times 10^n$

  l $0.0101 = 1.01 \times 10^n$

  m $100\,400 = 1.004 \times 10^n$

  n $0.00902 = 9.02 \times 10^n$

5 Write these in standard form.

  a 6 470 000 correct to 2 significant figures

  b 321 000 correct to 2 significant figures

  c 27 800 000 correct to 2 significant figures

  d 1 342 000 correct to 3 significant figures

  e 0.00439 correct to 2 significant figures

  f 0.001662 correct to 3 significant figures

  g 0.008219 correct to 3 significant figures

  h 0.000557 correct to 2 significant figures

  i 0.000027 correct to 1 significant figure

  j 7 210 000 correct to 1 significant figure

  k 917 000 correct to 1 significant figure

6 Arrange these numbers in order of size with the smallest first.

  a 3×10^4 2×10^7 9×10^{-6} 8×10^{-3}

  b 6×10^8 7×10^{-5} 1×10^9 3×10^{-2}

  c 6.4×10^5 8.1×10^{-1} 4.4×10^3 1.2×10^{-2}

  d 9.8×10^{-3} 2.5×10^6 3.75×10^4 7.6×10^{-7}

  e 0.005   3500 3.7×10^2 5×10^{-5}

  f 2×10^4 200 000   0.002 2×10^{-2}

  g 1.5×10^3 15 000   0.15 1.5×10^{-3}

  h 7.6×10^{-4} 0.0076   0.076 7.6×10^{-6}

The laws of indices

7 Write these volumes in litres in standard form.

a 7 cm^3	b 9 cm^3	c 35 cm^3	d 78 cm^3
e 125 cm^3	f 240 cm^3	g 1 cm^3	h 0.5 cm^3
i 0.45 cm^3	j 0.06 cm^3	k 0.075 cm^3	l 0.008 cm^3

8 Write these masses in kilograms in standard form.

a 4 g	b 2 g	c 27 g	d 89 g
e 340 g	f 575 g	g 1 g	h 0.5 g
i 0.72 g	j 0.03 g	k 0.005 g	l 0.0001 g

9 Write these distances in kilometres in standard form.

a 5 m	b 2 m	c 34 m	d 0.6 m
e 0.25 m	f 0.17 m	g 0.08 m	h 0.009 m
i 24 cm	j 37 cm	k 8 cm	l 6 cm
m 0.5 cm	n 0.25 cm	o 7 mm	p 12 mm
q 35 mm	r 0.5 mm	s 0.34 mm	t 0.04 mm

10 Calculate these using standard form.

Example
$$0.00034 \times 0.002 = 3.4 \times 10^{-4} \times 2 \times 10^{-3}$$
$$= 6.8 \times 10^{-7}$$
$$= 0.00000068$$

a 0.00023×0.003	b 0.004×0.00006
c 0.0007×0.05	d 0.042×0.002
e 0.008×0.023	f 0.00004×0.6
g 4000×2000	h $2300 \times 50\,000$
i $13\,000 \times 300$	j $320\,000 \times 0.004$
k $51\,000 \times 0.006$	l $0.042 \times 70\,000$
m $230\,000 \times 0.000003$	n 0.0006×500
o 1200×0.000008	p 4000×0.0009
q $0.0000005 \times 200\,000$	r $0.0000001 \times 1\,000\,000$

11 Multiply each pair of numbers together and give the answer in standard form.

a $6 \times 10^5, \; 2 \times 10^3$	b $3 \times 10^4, \; 5 \times 10^6$
c $7 \times 10^6, \; 3 \times 10^2$	d $8 \times 10^2, \; 5 \times 10^5$
e $4 \times 10^8, \; 6 \times 10^{-3}$	f $9 \times 10^7, \; 2 \times 10^{-4}$
g $6 \times 10^{11}, \; 5 \times 10^{-5}$	h $8.5 \times 10^{12}, \; 2 \times 10^{-3}$
i $2.5 \times 10^3, \; 6 \times 10^{-7}$	j $6.5 \times 10^4, \; 2 \times 10^{-9}$
k $7.2 \times 10^6, \; 5 \times 10^{-11}$	l $5 \times 10^{-2}, \; 3.4 \times 10^{-4}$
m $8 \times 10^{-3}, \; 4 \times 10^{-4}$	n $6.5 \times 10^{-7}, \; 8 \times 10^{-3}$

12 Use square tables and standard form to work out the value of these.

Example
$$(35\,600)^2 = (3.56 \times 10^4)^2$$
$$= 12.7 \times 10^8$$
$$= 1\,270\,000\,000$$

a 6250^2	b 2430^2	c $12\,800^2$	d 678^2
e 295^2	f $33\,400^2$	g 0.0843^2	h 0.0225^2
i 0.00342^2	j 0.175^2	k 0.905^2	l 0.00776^2

The laws of indices

13 Work these using standard form.

a $\dfrac{0.008}{400}$ b $\dfrac{0.06}{2000}$ c $\dfrac{0.009}{3000}$ d $\dfrac{0.08}{200}$

e $\dfrac{0.4}{2000}$ f $\dfrac{600}{30\,000}$ g $\dfrac{600}{0.003}$ h $\dfrac{15\,000}{0.05}$

i $\dfrac{240\,000}{0.06}$ j $\dfrac{180}{0.009}$

14 Work these and give the answers in standard form.

a $\dfrac{7.2 \times 10^6}{2 \times 10^3}$ b $\dfrac{7.2 \times 10^6}{9 \times 10^3}$ c $\dfrac{6.4 \times 10^5}{4 \times 10^2}$ d $\dfrac{6.4 \times 10^5}{8 \times 10^2}$

e $\dfrac{3.6 \times 10^{-2}}{2 \times 10^4}$ f $\dfrac{3.6 \times 10^{-2}}{4 \times 10^4}$ g $\dfrac{2.4 \times 10^{-3}}{6 \times 10^5}$ h $\dfrac{8.1 \times 10^{-7}}{9 \times 10^3}$

i $\dfrac{4.8 \times 10^3}{8 \times 10^{-4}}$ j $\dfrac{6.2 \times 10^6}{2 \times 10^{-3}}$ k $\dfrac{2.7 \times 10^{11}}{3 \times 10^{-4}}$ l $\dfrac{9 \times 10^3}{2 \times 10^{-5}}$

15 Give answers to these in standard form.

a Find the area of a square in mm^2 if each side is $12\,000$ mm long.

b Find the volume of a cube in mm^3 if each edge is $5\,000$ mm long.

c Find the value of E if $E = mc^2$ and $c = 3 \times 10^6$ and $m = 0.00005$.

d Find the value of J if $J = Ri^2t$ and $R = 40\,000$, $i = 0.003$, $t = 200$.

e Find the area of a circle in mm^2 if π is 3 and the radius is $20\,000$ mm.

16 a Light travels at a speed of 3×10^8 m/s. Find how far it travels in one hour. Give the answer in standard form
(i) in metres (ii) in kilometres.

b Sound travels at a speed of 3.3×10^5 cm/s in still air at sea-level. Find how far it travels in one minute. Give the answer in standard form correct to one significant figure
(i) in centimetres (ii) in metres.

Powers of two and three

Copy and complete these two tables.

Power of 2	Its value
0	$2^0 = 1$
1	$2^1 = 2$
2	$2^2 = 4$
3	$2^3 = 8$
4	$2^4 =$
5	$2^5 =$
6	
7	
8	
9	
10	
11	
12	

Power of 3	Its value
0	$3^0 = 1$
1	$3^1 = 3$
2	$3^2 = 9$
3	$3^3 = 27$
4	$3^4 = 81$
5	$3^5 =$
6	
7	
8	
9	
10	
11	
12	

Work these multiplications and divisions, by following the methods shown.

Examples **a** $16 \times 128 = 2^4 \times 2^7 = 2^{11} = 2048$

b $\dfrac{6561}{27} = \dfrac{3^8}{3^3} = 3^5 = 729$

1 32×64

2 16×256

3 128×4

4 32×128

5 81×243

6 27×2187

7 243×729

8 27×6561

9 729×9

10 6561×81

11 $\dfrac{256}{32}$

12 $\dfrac{1024}{16}$

13 $\dfrac{2048}{128}$

14 $\dfrac{512}{64}$

15 $\dfrac{729}{81}$

16 $\dfrac{243}{27}$

17 $\dfrac{2187}{729}$

18 $\dfrac{19\,683}{729}$

19 $\dfrac{59\,049}{6561}$

20 $\dfrac{177\,147}{2187}$

21 $\dfrac{531\,441}{243}$

22 $\dfrac{32 \times 64}{128}$

23 $\dfrac{128 \times 32}{16}$

24 $\dfrac{512 \times 8}{128}$

25 $\dfrac{1024 \times 16}{128}$

26 $\dfrac{6561 \times 81}{243}$

27 $\dfrac{19\,683 \times 27}{729}$

28 $\dfrac{243 \times 729}{19\,683}$

29 $\dfrac{19\,683 \times 81}{243}$

30 $\dfrac{1024 \times 256}{64 \times 4096}$

31 $(243)^2$

32 $(64)^2$

33 $(729)^2$

34 $(128)^2$

35 $\dfrac{6561 \times 1024}{256 \times 2187}$

36 $\dfrac{177\,147 \times 1024}{531\,441 \times 4096}$

The slide-rule

Part 1 Using powers of two

1

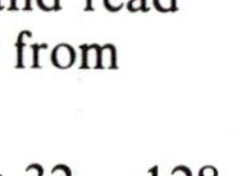

Take two strips of card a little over 10 cm long.

Draw marks 1 cm apart on the bottom edge **C** and top edge **D**.

Label the marks with the powers of 2 as shown, and fill in the missing powers.

2 To calculate 4×32, place **C** 1 above **D** 4, find **C** 32 and read the answer from edge **D**.

So that $4 \times 32 = 128$.

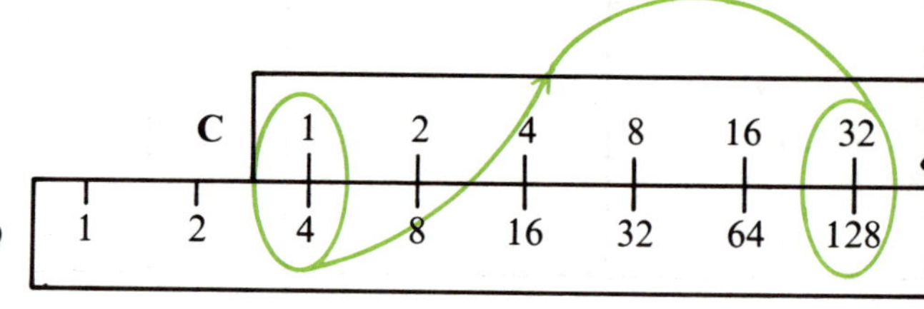

Work the following multiplications in the same way.

a 8×16	**b** 4×64	**c** 16×32	**d** 32×4
e 128×2	**f** 8×32	**g** 16×16	**h** 8×128
i 32×32	**j** 8×8	**k** 64×8	**l** 8×64
m 128×4	**n** 256×2	**o** 16×4	

3 To calculate $\dfrac{128}{16}$, place **C** 16 above **D** 128, find **C** 1 and read the answer from edge **D**.

So that $\dfrac{128}{16} = 8$.

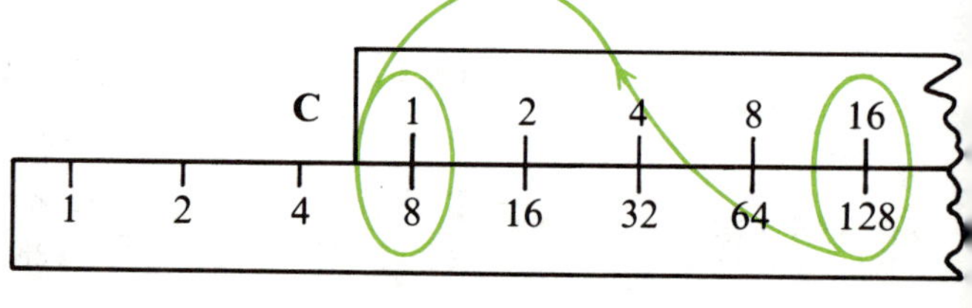

Work these divisions in the same way.

a $\dfrac{128}{8}$	**b** $\dfrac{256}{32}$	**c** $\dfrac{512}{16}$	**d** $\dfrac{256}{8}$
e $\dfrac{1024}{64}$	**f** $\dfrac{512}{4}$	**g** $\dfrac{128}{2}$	**h** $\dfrac{512}{128}$
i $\dfrac{256}{32}$	**j** $\dfrac{128}{4}$	**k** $\dfrac{1024}{8}$	**l** $\dfrac{512}{2}$
m $\dfrac{64}{16}$	**n** $\dfrac{256}{2}$	**o** $\dfrac{1024}{4}$	

The slide-rule

4 Work these with the two strips of card.

a $4 \times 16 \times 8$ b $32 \times 4 \times 8$ c $2 \times 128 \times 4$ d $16 \times 16 \times 2$

e $4 \times 32 \times 4$ f $8 \times 64 \times 2$ g $\dfrac{16 \times 32}{64}$ h $\dfrac{4 \times 128}{32}$

i $\dfrac{128 \times 8}{256}$ j $\dfrac{16 \times 16}{64}$ k $\dfrac{256 \times 4}{32}$ l $\dfrac{128 \times 8}{64}$

m $\dfrac{32 \times 8}{128}$ n $\dfrac{16 \times 8}{128}$ o $\dfrac{2 \times 256}{64}$ p $\dfrac{8 \times 32}{16}$

Part 2 Reading the scales

Three different scales are used on commercial slide-rules. Care must be taken to distinguish between them.

Write the decimal numbers to which the lettered arrows point.

1 

2 

3 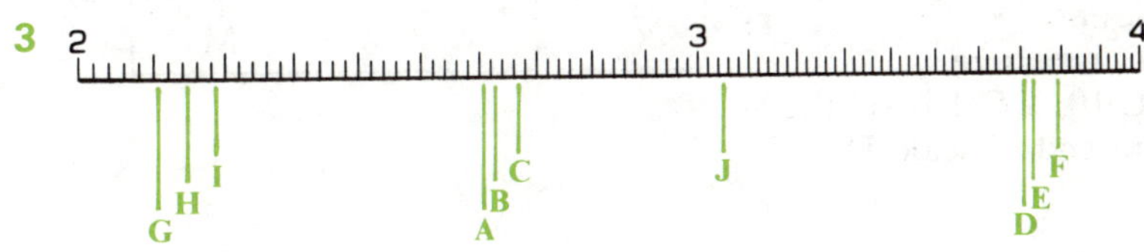

4 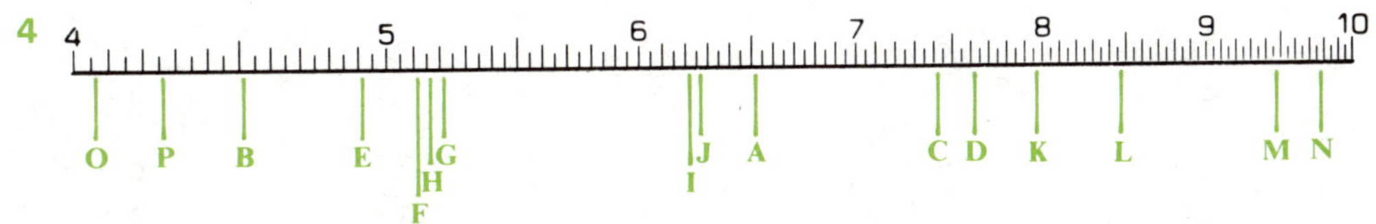

The slide-rule

Part 3 Using the slide-rule

1 To calculate M × N, position the numbers as shown on this diagram.

The answer is found under N on scale **D**.

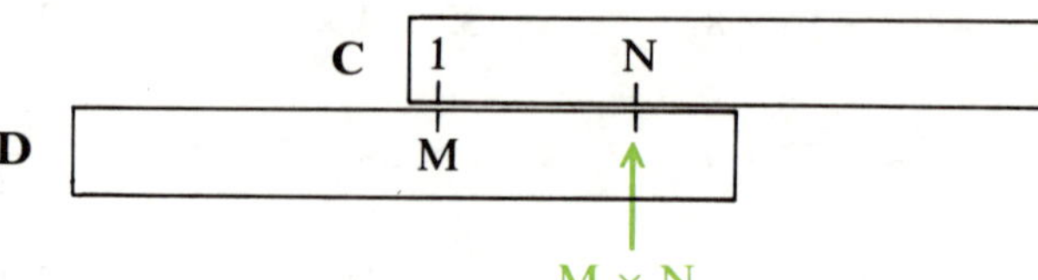

Work these.

a	1.5 × 4	b	1.2 × 5	c	2 × 1.5	d	1.8 × 5
e	1.4 × 5	f	1.6 × 5	g	1.5 × 5	h	1.2 × 7.5
i	2.5 × 4	j	2.5 × 3	k	2.4 × 2.5	l	3.2 × 2.5
m	2.4 × 3	n	4.4 × 1.5	o	3.2 × 1.5	p	2.6 × 3
q	4.2 × 1.5	r	3.5 × 1.8	s	2.6 × 3.5	t	5.2 × 1.5
u	2.2 × 3.5	v	4.6 × 1.5	w	3.5 × 2.8	x	1.75 × 2.8

2 In some multiplications, the answer is off the right-hand end of the **D** scale.

In this case, put **C** 10 rather than **C** 1 above the number M.

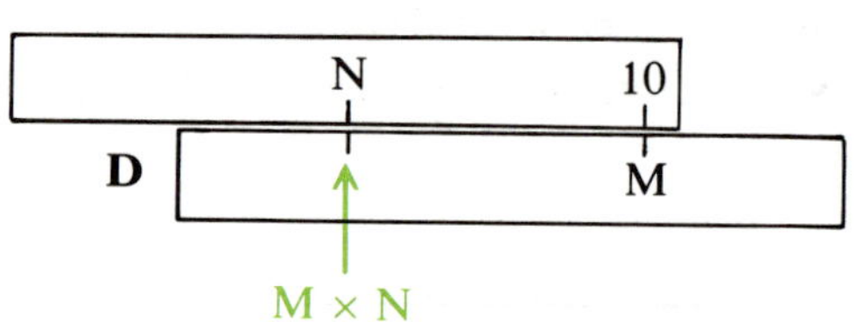

The answer is still found under N on scale **D**; but take care to place its decimal point correctly.

Work these.

a	7.5 × 4	b	8.5 × 2	c	9.5 × 2	d	8 × 2
e	5.5 × 2	f	6.5 × 8	g	6 × 7.5	h	7.5 × 2.6
i	8.1 × 2.1	j	7.1 × 3.1	k	6.9 × 4.2	l	6.2 × 6.3
m	5.2 × 4.8	n	5.3 × 5.1	o	4.7 × 6.6	p	4.2 × 3.1
q	3.9 × 4.1	r	9.1 × 6.1	s	8.7 × 7.3	t	7.1 × 4.65

3 To calculate $\dfrac{M}{N}$, position the numbers as shown on this diagram.

The answer is found
either under **C** 1 as here,
or under **C** 10, if **C** 1 is off the left-hand end of scale **D**.

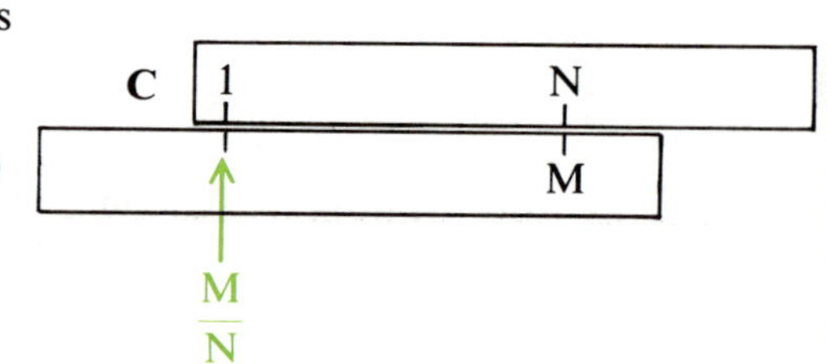

Work these.

a	$\dfrac{6}{4}$	b	$\dfrac{8}{5}$	c	$\dfrac{7}{5}$	d	$\dfrac{5}{2}$	e	$\dfrac{5}{4}$
f	$\dfrac{7}{4}$	g	$\dfrac{6.5}{5}$	h	$\dfrac{4.5}{3}$	i	$\dfrac{9}{7.5}$	j	$\dfrac{6}{2.5}$
k	$\dfrac{2.4}{1.6}$	l	$\dfrac{4.5}{2.5}$	m	$\dfrac{6.5}{2.5}$	n	$\dfrac{7.7}{3.5}$	o	$\dfrac{6.5}{2.1}$

The slide-rule

p $\dfrac{8.9}{2.4}$ q $\dfrac{9.1}{3.5}$ r $\dfrac{17}{5}$ s $\dfrac{18}{4}$ t $\dfrac{21}{3.5}$

u $\dfrac{45}{7.9}$ v $\dfrac{53}{6.8}$ w $\dfrac{49}{8.3}$ x $\dfrac{29}{6.3}$ y $\dfrac{37}{4.3}$

z $\dfrac{8.9}{3.3}$

4 Work these on a slide-rule.

a $\dfrac{1.5 \times 3.8}{3.0}$ b $\dfrac{1.2 \times 4.5}{1.8}$ c $\dfrac{1.8 \times 2.5}{3.6}$

d $\dfrac{2.5 \times 3.6}{1.5}$ e $\dfrac{3.3 \times 1.25}{1.8}$ f $\dfrac{1.65 \times 3.4}{2.8}$

g $\dfrac{7.9 \times 1.1}{6.2}$ h $\dfrac{3.9 \times 2.1}{4.1}$ i $\dfrac{1.5 \times 4.2}{3.5}$

j $\dfrac{8.1}{1.45 \times 3.1}$ k $\dfrac{9.1}{2.6 \times 2.5}$ l $\dfrac{5.4}{1.12 \times 3.3}$

m $\dfrac{7.8}{2.58 \times 1.9}$ n $\dfrac{8.7}{3.6 \times 2.14}$ o $\dfrac{3.7}{2.4 \times 3.5}$

p $\dfrac{5.4 \times 1.5}{8.9}$ q $\dfrac{3.2 \times 1.75}{6.3}$ r $\dfrac{1.23 \times 3.9}{5.4}$

s $\dfrac{2.5 \times 1.76}{6.2}$ t $\dfrac{12.4 \times 6.25}{8.25}$

Part 4 Ratio problems

In some types of problem, a slide-rule is even quicker than a pocket calculator.

1 A teacher marks an examination which has a maximum possible mark of 80.
Convert the marks of his twelve pupils into percentages, by using a slide-rule on which **C** 8 is above **D** 10.
Copy and complete this table.

Mark (out of 80)	40	20	60	44	36	72	64	56	68	76	16	12
Percentage												

2 5 miles is almost the same distance as 8 km.
Place **C** 5 above **D** 8 and use the slide-rule to convert these distances from miles to kilometres.
Copy and complete this table.

Miles	2	3	4	1.5	3.5	4.5	5.5	6	1.25	1.75
Kilometres										

The slide-rule

3 £1 can be exchanged for 15 Danish kroner.

Place **C** 1 above **D** 15 and convert this British currency into kroner.

Copy and complete this table.

£	2	4	5	1·40	1·60	2·20	2·40	2·80	4·20	4·80	5·20	6·40
Kroner												

4 A history examination has a maximum possible mark of 125.

Place **C** 1 above **D** 125, and convert these ten results into percentages.

Copy and complete this table.

Mark (out of 125)	40	50	70	80	65	95	45	$77\frac{1}{2}$	$82\frac{1}{2}$	$97\frac{1}{2}$
Percentage										

5 1 cm³ of iron has a mass of 7.2 grams.

Place **C** 10 above **D** 72, and find the mass of each of the ten pieces of iron with the volumes given in the table.

Copy and complete this table.

Volume, cm³	0.5	0.25	0.75	0.57	0.15	0.21	0.26	0.555	0.625	0.945
Mass, grams										

Computation

Introduction Powers of ten

Logarithm tables are used to write any number between 1 and 10 as a power of ten e.g. $2.56 = 10^{0.408}$.

Examples

a Multiplication
$$2.56 \times 1.83 = 10^{0.408} \times 10^{0.262}$$
$$= 10^{0.670}$$
$$= 4.68$$

b Division $\dfrac{8.64}{3.25} = \dfrac{10^{0.937}}{10^{0.512}} = 10^{0.425} = 2.66$

Work these using powers of ten. Set out the work as in the above examples. Check the answers to nos. 1–5 by long multiplication. All the answers could also be checked with a calculator.

1 2.31×3.12

2 4.04×1.38

3 1.27×7.11

4 3.25×2.39

5 6.12×1.15

6 $\dfrac{9.04}{4.17}$

7 $\dfrac{7.87}{4.12}$

8 $\dfrac{5.02}{3.56}$

9 $\dfrac{8.40}{5.51}$

10 $\dfrac{9.86}{1.03}$

Part 1 Multiplication of numbers less than 10

Example $3.21 \times 1.43 = 4.59$

number	log
3.21	0.507
× 1.43	0.155 +
4.59	0.662

Work these.

1 4.04×1.45

2 2.05×3.84

3 4.50×1.64

4 1.79×4.08

5 3.28×1.71

6 1.43×5.58

7 2.85×2.67

8 2.28×3.14

9 1.69×4.44

10 4.38×1.89

11 1.31×6.52

12 2.66×2.47

13 $2.14 \times 3.25 \times 1.36$

14 $2.72 \times 1.76 \times 1.94$

15 $1.61 \times 2.66 \times 1.98$

16 $2.04 \times 1.25 \times 3.11$

17 $3.42 \times 2.12 \times 1.33$

18 $1.64 \times 2.11 \times 1.75 \times 1.27$

19 A hearth rug measures 2.15 m by 1.24 m. Find its area in m^2.

20 Find the area of a table top which is 3.64 m long and 1.85 m wide.

21 A sheet of glass for a shop window is 2.15 m by 1.25 m.
   a Find its area in m^2.
   b If 1 m^2 costs £3·35, find the cost of the glass.

22 A garden shed has the shape of a cuboid and is 2.95 m long, 1.75 m wide and 1.85 m high. Find the volume of the shed in m^3.

Computation

23 The cold water tank in a house has a square base 1.64 m by 1.64 m and is 1.36 m high. Find the capacity of the tank.

24 A cube of metal has an edge length of 1.25 cm.
   a Find its volume.
   b Find its mass, if 1 cm³ has a mass of 4.5 grams.

25 If £1 is worth $1.94, how many dollars would you get for
   a £4   **b** £4·26?

26 If £1 is worth 3.94 DM, how many German marks would you get for
   a £2   **b** £2·38?

27 1 cm³ of aluminium has a mass of 2.70 grams. What is the mass of
   a 3 cm³   **b** 3.27 cm³?

28 1 cm³ of sandstone has a mass of 2.26 grams. Find the mass of
   a 4 cm³   **b** 4.18 cm³.

29 Find **a** 1% of £324   **b** 2.75% of £324.

30 An automatic cutting machine cuts metal strips supposedly 412 cm long. If the strips are inaccurate by more than 1.82% of this, then they are rejected. Find
   a 1% of 412 cm
   b 1.82% of 412 cm
   c the greatest and the least acceptable lengths of the strips.

Part 2 Division of numbers less than 10

Example $\dfrac{8.21}{3.32} = 2.47$

number	log
8.21	0.914
÷ 3.32	0.521 −
2.47	0.393

Work these.

1 $\dfrac{8.61}{7.98}$

2 $\dfrac{7.71}{4.43}$

3 $\dfrac{7.55}{2.05}$

4 $\dfrac{7.21}{1.71}$

5 $\dfrac{6.01}{2.52}$

6 $\dfrac{8.11}{2.44}$

7 $\dfrac{7.26}{3.44}$

8 $\dfrac{7.78}{1.88}$

9 $\dfrac{6.87}{3.08}$

10 $\dfrac{7.43}{2.37}$

11 $\dfrac{6.92}{2.70}$

12 $\dfrac{5.62}{2.41}$

13 $\dfrac{4.74}{2.56}$

14 $\dfrac{3.65 \times 2.46}{8.52}$

15 $\dfrac{4.27 \times 1.89}{6.43}$

16 $\dfrac{2.03 \times 4.69}{3.11}$

17 $\dfrac{9.05}{2.25 \times 1.58}$

18 $\dfrac{7.94}{3.28 \times 1.87}$

19 $\dfrac{8.66}{4.03 \times 1.52}$

20 $\dfrac{3.45 \times 2.87}{5.03 \times 1.12}$

Computation

21 A sheet of hardboard has an area of 8.75 m². If its length is 3.64 m, find its width.

22 A carpet of area 9.2 m² just fits a room of width 2.65 m. Find the length of the room.

23 Find how many dollars can be changed for £1, if
   a   £3 = \$5.94   b   £3·45 = \$6.83.

24 Find how many German marks can be changed for £1, if
   a   £2 = 7.84 DM   b   £2·35 = 9.21 DM.

25 If 1.24 cm³ of iron have a mass of 8.93 grams, find the mass of 1 cm³.

26 If a diamond has a mass of 8.27 grams and a volume of 2.35 cm³, what will be the mass of a 1 cm³ diamond?

27 7.25 m² of carpet are bought for £8·99. What is the cost of 1 m²?

28 A cuboid has a volume of 7.94 cm³. If its length is 3.15 cm and its width is 1.2 cm, how high is it?

29 A water tank can hold a maximum of 3.83 m³ of water. If it has a square base of side 1.25 m, find its height.

30 A small block of iron is 3.85 cm long and 1.2 cm high. If its volume is 6.47 cm³, find the width of the block.

Part 3 Multiplication and Division with numbers greater than ten

Introduction   Copy and complete these two tables.

Number	As a power of ten	Its logarithm
1	10⁰	0.000
10	10¹	1.000
100	10²	2.000
1000		
10 000		
100 000		
1 000 000		

Number	Its logarithm
6.45	
64.5	
645	
6450	
64 500	
645 000	
6 450 000	

Work these.

1	6.03 × 8.61	6	5.62 × 7.84 × 6.73	11	$\dfrac{49.7}{6.62}$
2	4.48 × 5.59	7	2.74 × 8.92 × 7.60	12	$\dfrac{54.8}{8.47}$
3	7.44 × 4.87	8	32.6 × 23.5	13	$\dfrac{28.7}{5.57}$
4	3.38 × 8.07	9	89.6 × 40.3	14	$\dfrac{674}{9.25}$
5	5.47 × 9.31	10	327 × 16.5	15	$\dfrac{204}{8.45}$

Computation

16 $\dfrac{589}{73.5}$		**23** $\dfrac{18.7 \times 46.0}{274}$		**30** $\dfrac{8.46 \times 7.02}{2.85 \times 6.33}$	
17 $\dfrac{4350}{275}$		**24** $\dfrac{353 \times 50.2}{890}$		**31** $\dfrac{5.05 \times 43.6}{28.9 \times 3.14}$	
18 $\dfrac{6870}{755}$		**25** $\dfrac{79.2 \times 406}{2470}$		**32** $\dfrac{67.4 \times 424}{280 \times 36.2}$	
19 $\dfrac{7450}{32}$		**26** $\dfrac{8.92}{3.72 \times 1.47}$		**33** $\dfrac{755 \times 39.0}{15.8 \times 56.5}$	
20 $\dfrac{276}{49}$		**27** $\dfrac{68.4}{13.2 \times 3.06}$		**34** $\dfrac{6.87 \times 30.4 \times 215}{3240}$	
21 $\dfrac{6.72 \times 8.43}{32.1}$		**28** $\dfrac{428}{27.5 \times 12.2}$		**35** $\dfrac{8.42 \times 10.9 \times 246}{507 \times 4.45}$	
22 $\dfrac{5.04 \times 27.3}{68.4}$		**29** $\dfrac{960}{14.3 \times 1.77}$			

Part 4 Problems

1. A gardener plants his vegetables in a rectangular plot 18.5 m by 7.6 m. Find the area of this plot.

2. A cupboard door has an area of 9280 cm². If the height of the door is 145 cm, find its width.

3. A living-room is 8.75 m long and 4.8 m wide.
   a. Find the area of the floor of the room.
   b. If carpet is bought costing £5·60 per m², find the cost of fitting carpet in this room (to the nearest £).

4. A rectangular field is 328 m long and 142 m wide.
   a. Find its area in square metres.
   b. It is sown with corn so that on average there are 375 grains per m². Find the number of grains of corn sown in the field.

5. a. The A38 main road runs from Bodmin in Cornwall to Derby, a distance of 272 miles. If 1 mile = 1.61 km, find this distance in kilometres.
   b. The A9 main road runs from Edinburgh to John o' Groats, a distance of 288 miles. How far is this in kilometres?

6. a. It is 557 km from Oslo to Trondheim in Norway. If 1 mile = 1.61 km, find the distance in miles.
   b. From Calais on the English Channel to Rome is 1680 km. How far is it in miles?

7. a. I fill the petrol-tank of my car with 8.55 gallons. If 1 gallon = 4.54 litres, how many litres have I put in?
   b. My house has oil-fired central heating and the oil tank will hold 875 galls. How many litres does it hold?

8. a. A housewife buys 23.8 litres of cooking oil. How many gallons is this?
   b. A plastic bucket holds 15.9 litres. How many gallons does it hold?

9. If I cycle for 8.75 hours at an average speed of 13.5 mph, how far will I travel?

Computation

10 The travelling time (excluding overnight stops) for a bus journey from London to Athens is 56 hours. If the average speed during the day is 35.5 mph, find the distance (in miles) from London to Athens.

11 Leeds to London is 195 miles and Mr Earnshaw made the journey by car. Find his average speed (in mph), if he took
a 3 h   b 3.75 h.

12 The distance from London to Nairobi in Kenya is 4450 miles. Find the average speed (in mph) of an aeroplane, if the journey took
a 10 h   b 13.5 h.

13 Mr Swift has a sports car in which he travelled from Bristol to Glasgow in 5.5 hours at an average speed of 66.5 mph. His friend, Mr Longer, in his minibus took 6.5 hours for the same trip. Find
a how far it is from Bristol to Glasgow
b Mr Longer's average speed.

14 If you went to France for your holiday and the exchange rate was £1 = 8.45 fr., how many francs would you get for
a £2   b £54·50?

15 If you returned from a holiday in France with 262 fr. left over and you changed them for £31·40, how many francs did the new exchange rate give for £1?

16 Emma Robson went to Italy on a school trip when the exchange rate was £1 = 1630 lire. How many lire did she get for
a £2   b £46·50?

When she came back from holiday, the exchange rate had altered, so that for the 8550 lire she had left, she received £5·12.
c How many lire did she get for £1 at this new rate?

17 Olaf Erikson lives in Bergen, Norway. He came by ship to Newcastle-upon-Tyne where he went on a shopping spree. The exchange rate was £1 = 10.6 kr. and he changed 2450 kr. into pounds £.
a How many British pounds was he given?

When he left for home two days later, he had only £4·25 left. The exchange rate had altered slightly and he received 43·8 kr. in exchange.
b What did the new exchange rate give him for £1?

18 A tea chest 84 cm high has a base 54 cm by 48 cm. It is placed into the back of a furniture lorry which is 4.5 m long, 2.85 m across and 2.5 m high. Find the volume of
a the chest (in cm^3)   b the back of the lorry (in m^3).

19 A steel girder is 127 cm long, 8.5 cm across and 5.5 cm deep. If 1 cm^3 of steel has a mass of 7.5 grams, find
a the volume of the girder
b its mass in grams   c its mass in kg.

20 The volume of a room is 168 m^3. If its length is 8.75 m and its width is 6.35 m, what is its height?

21 A paving slab, 85 cm long and 28 cm across, has a volume of 8.33 litres. Find
a its volume in cm^3 b its thickness.

22 A block of metal has a mass of 6.32 kg, and is 5.6 cm wide and 4.8 cm deep. Find
a its mass in grams
b the volume of the block, if 1 cm^3 of metal has a mass of 8.55 grams
c the length of the block.

Computation

Part 5 Powers of numbers

Example $5.84^3 = 198$ *Either*

number	log
5.84	0.766
5.84	0.766
× 5.84	0.766 +
198	2.298

or

number	log
5.84	0.766
	× 3
198	2.298

Work these.

1 2.06^2

2 2.85^2

3 3.08^2

4 4.25^3

5 2.97^4

6 3.57^4

7 5.21^4

8 13.7^2

9 25.4^2

10 18.5^3

11 7.45^4

12 3.48^6

13 2.38^8

14 $(1.35 \times 12.7)^3$

15 $(26.3 \times 1.33)^2$

16 $(2.07 \times 1.83)^7$

17 $\left(\dfrac{86.4}{3.72}\right)^3$

18 $\left(\dfrac{53.9}{21.5}\right)^5$

19 $\left(\dfrac{279}{97.2}\right)^2$

20 $\left(\dfrac{3750}{849}\right)^3$

21 A cube of metal has edges 15.6 cm long. Find the volume of the cube in cm^3.

22 A petrol tank at a garage has the shape of a cube of edge 3.8 metres.
Find the volume of petrol in a a full tank
b a half-empty tank.

23 If prices are increasing by 15% each year due to inflation, then the cost C in T years time of an article which now costs C_o is given by the formula
$$C = C_o \times 1.15^T.$$
Find C when a $C_o = £25$, $T = 4$
b $C_o = £25$, $T = 8$.

24 If inflation of 5% per annum increases the present cost C_o of an article to a higher cost C in T years, then
$$C = C_o \times 1.05^T.$$
Find C when a $C_o = £25$, $T = 4$
b $C_o = £25$, $T = 8$.

25 If you deposit $£P$ in a bank account where it earns 12% per annum, then in T years time you will have $£Q$ in your account, where
$$Q = P \times 1.12^T.$$
Find Q when a $P = 68.5$, $T = 3$
b $P = 740$, $T = 4$.

26 On January 1st you save 1p; on January 2nd you save 2p; on January 3rd you save 4p; and so on, doubling the amount you save every day.
After D days, you will have saved $2^D - 1$ pence.
What are your savings in £ a on January 21st
b on January 31st?

Computation

Part 6 Roots of numbers

Example $\sqrt[3]{75.1} = 4.22$

number	log
75.1	1.876
	$\div$ 3
4.22	0.625

Work the following.

1 $\sqrt{9.24}$

2 $\sqrt{4.11}$

3 $\sqrt{52.7}$

4 $\sqrt[3]{71.5}$

5 $\sqrt[3]{8.99}$

6 $\sqrt[3]{17.0}$

7 $\sqrt[4]{305}$

8 $\sqrt[4]{717}$

9 $\sqrt[5]{603}$

10 $\sqrt[5]{91.4}$

11 $\sqrt[8]{1940}$

12 $\sqrt[6]{3860}$

13 $\sqrt[3]{14.7 \times 21.3}$

14 $\sqrt{9.84 \times 86.2}$

15 $\sqrt{242 \times 13.3}$

16 $\sqrt[3]{\dfrac{863}{31.4}}$

17 $\sqrt[4]{\dfrac{207}{3.7}}$

18 $\sqrt[5]{\dfrac{4370}{11.5}}$

19 $\sqrt{\dfrac{11.2}{9.8}}$

20 $\sqrt[3]{8.4 \times 42.7 \times 123}$

21 A wooden cube has a volume of 325 cm³. Calculate the length of one of its edges.

22 A cube-shaped tank holds 41.5 m³ of acid when full. How long is each edge of the tank?

23 A newly built city square has an area of 9120 m². How long is each side of the square?

24 The current I flowing through a resistor can be found from the formula $I = \sqrt{\dfrac{w}{R}}$.
Find I when **a** $w = 67.5,\ R = 2.7$
b $w = 842,\ R = 37.5$.

25 The length x in this pyramid can be found
from its volume V using the formula $x = \sqrt[3]{3V}$.
Find x when **a** $V = 21.3$ cm³
b $V = 592$ cm³.

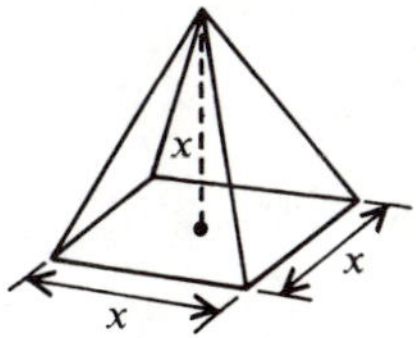

26 The radius r of a cylinder is given by the formula

$r = \sqrt{\dfrac{V}{\pi h}}$, where V is its volume, h is its height and $\pi = 3.14$.

Calculate the radius r when **a** $V = 86.5$ cm³, $h = 3.0$ cm
b $V = 250$ cm³, $h = 5.0$ cm.

Computation

Part 7 A mixture

Work these using logarithms. All the answers could be checked with a calculator.

1   $\dfrac{12.8 \times 87.3}{52.4}$

2   $\dfrac{8.52 \times 124}{47.9}$

3   $\dfrac{14.9 \times 89.4}{225}$

4   $\dfrac{89.4}{14.6 \times 3.65}$

5   $\dfrac{267}{124 \times 1.15}$

6   $\dfrac{3290}{807 \times 2.43}$

7   $\dfrac{42.8 \times 50.4}{8.82 \times 3.15}$

8   $\dfrac{7.62 \times 703}{252 \times 5.51}$

9   $\dfrac{8.43 \times 7.84 \times 424}{3040 \times 6.21}$

10   $(17.2)^2 \times 6.75$

11   $(20.4)^3 \times 1.16$

12   $\dfrac{(8.62)^4}{573}$

13   $\dfrac{(82.6)^2}{72.5}$

14   $\left(\dfrac{82.6}{72.5}\right)^2$

15   $\left(\dfrac{81.4 \times 27.2}{253}\right)^2$

16   $\left(\dfrac{19.7 \times 90.5}{39.2}\right)^3$

17   $\left(\dfrac{3640}{8.72 \times 39.6}\right)^2$

18   $\sqrt{29.7 \times 53.6}$

19   $\sqrt{2.07 \times 535}$

20   $\sqrt[3]{15.6 \times 82.4}$

21   $\sqrt[4]{92.7 \times 3.03}$

22   $\sqrt{\dfrac{64.9}{2.55}}$

23   $\sqrt{\dfrac{2096}{12.4}}$

24   $\sqrt[3]{\dfrac{8140}{5.27}}$

25   $\sqrt[4]{\dfrac{707}{2.40}}$

26   $\sqrt[3]{(14.7)^2}$

27   $\sqrt[3]{(2.94)^5}$

28   $\sqrt{(8.72)^3}$

29   $\sqrt[4]{\dfrac{(62.7)^2}{89.4}}$

30   $\sqrt{\dfrac{92.1 \times 3.67}{214}}$

31   $\sqrt{\dfrac{369 \times 52.8}{2090}}$

32   $\sqrt[3]{\dfrac{5120}{40.6 \times 1.97}}$

33   $\sqrt[3]{\dfrac{42.9 \times 8.07}{114 \times 2.13}}$

34   $\left(\dfrac{\sqrt[3]{92.4}}{2.67}\right)^2$

35   $\left(\dfrac{\sqrt[3]{365}}{1.67}\right)^4$

36   $\left(\dfrac{\sqrt{19.8}}{4.20}\right)^5$

37   $2.34^5 + 8.23^2$

38   $3.67^2 + 9.04^2$

39   $7.23^3 - 4.79^3$

40   $2.95^3 - \sqrt[3]{1250}$

41   $12.9^2 + \sqrt{984}$

42   $\left(\dfrac{76.5}{14.2}\right)^2 - 2.13^4$

43   $\dfrac{30.7}{3.80} - \sqrt[3]{100}$

44   $\dfrac{142}{19.7} + \sqrt[3]{10}$

45   $\sqrt{\dfrac{18.7}{1.12}} + 9.45$

Computation

46 Find the area of a rectangular lawn 6.85 m long and 4.6 m wide.

47 A rectangular field is 145 m wide and 159 m long. Find its area
    a in m^2 b in hectares.

48 Find a the area of a square of side 6.55 m
          b the length of side of a square of area 6.55 m^2
          c the area of a square of side 14.5 m
          d the length of side of a square of area 14.5 m^2.

49 A shoe box is 42.5 cm long, 15.5 cm wide and 12.5 cm high. Calculate its volume.

50 A railway waggon is 4.7 m long, 2.35 m across and 1.95 m deep.
    a Calculate the capacity of one waggon in m^3.
    b Calculate the capacity of a train with 146 waggons.
    c If the train is filled with lorry-loads of coal, each load having a volume of 12.5 m^3, how many loads are needed?

51 An oil tank is 2.4 m long, 1.65 m wide and 1.3 m high.
    a Calculate the volume of oil it will hold when full (i) in m^3 (ii) in litres.
    b If the oil is emptied into drums, each holding 6.5 litres, find how many drums it will fill.

52 Water flowing at a rate of 24.5 litres per minute, fills an empty water tank which is 3.2 m long, 1.45 m wide and 1.2 m high. Calculate
    a the volume of water in a full tank (i) in m^3 (ii) in litres
    b how long it will take to fill the tank.

53 A metal bar is 84 cm long, 2.3 cm wide and 1.4 cm high, and each cubic centimetre of the metal has a mass of 8.93 grams. Find
    a the volume of the bar   b its mass (i) in grams   (ii) in kg.

54 A room contains 6750 m^3 of air. If the floor of the room is 35.5 m long and 21.5 m wide, find
    a the area of the floor   b the height of the room.

55 A tank holds 5.3 m^3 of water and its base is 2.75 m long by 1.45 m wide. Calculate
    a the area of its base   b the depth of water in the tank.

56 a If 23.5 cm^3 of lead have a mass of 268 grams, find the mass of 1 cm^3 of lead.
    b If 65.5 cm^3 of iron have a mass of 515 grams, find the mass of 1 cm^3 of iron.
    c What is the density of tin (in grams per cm^3), if 8.65 cm^3 of tin have a mass of 63 grams?

57 A rectangular sheet of metal is 56.5 cm long, 46.5 cm wide and 2 cm thick. If it has a mass of 64.1 kg, find
    a its mass in grams   b the volume of metal in the sheet in cm^3
    c the density of the metal (in grams per cm^3).

58 If £1 = 2.32 US dollars, find
    a how many dollars there are in £26·50
    b how many £ there are in 230 US dollars.

59 If £1 = 3.95 DM, find
    a how many DM there are in £37·50
    b how many £ there are in 57.40 DM.

60 If £1 = 1740 lire, find
    a how many lire there are in £38·60
    b how many £ there are in 8590 lire.

Computation

61 Take 1 metre as 39.4 inches and find
   a how many inches there are in 4.62 m
   b how many inches there are in 30.4 m
   c how many metres there are in 100 inches.

62 Use the formula $A = \pi r^2$ with π as 3.14 to find the area of the circles with these radii.
   a 6.75 cm   **b** 12.4 cm   **c** 245 cm

63 Use the formula $C = 2\pi r$ with π as 3.14 to find the circumference of the circles with these radii.
   a 6.75 cm   **b** 12.4 cm   **c** 245 cm

64 Use the formula $V = \pi r^2 h$ to find the volume of a cylinder when
   a $r = 8.4$ cm, $h = 39.6$ cm   **b** $r = 15.7$ cm, $h = 2.6$ cm.

65 A works canteen has 14.5 litres of milk and it uses 3.5 cm^3 in each cup of tea. Find
   a how many cm^3 of milk the canteen has
   b how many cups of tea can be made with it.

66 Find **a** the length of side of a square whose area is 74 cm^2
        b the length of one edge of a cube whose volume is 74 cm^3
        c the length of side of a square whose area is 426 cm^2
        d the length of one edge of a cube whose volume is 1620 cm^3.

67 465 metal ingots, each with a volume of 185 cm^3, are melted down to make 1250 cubes. Find
   a the volume of the metal used   **b** the volume of one cube
   c the length of edge of each cube.

68 If $R = \dfrac{R_1 R_2}{R_1 + R_2}$, find R when
   a $R_1 = 8.5$, $R_2 = 6.2$ **b** $R_1 = 28.3$, $R_2 = 14.5$.

69 If $\dfrac{PV}{T} = k$, find the value of k when
   a $P = 14.2$, $V = 33.5$, $T = 295$ **b** $P = 26.7$, $V = 18.6$, $T = 311$.

70 If $E = \dfrac{1}{2}mv^2$, find the value of E when
   a $m = 64$, $v = 12.6$ **b** $m = 2.04$, $v = 13.4$.

71 If $A = 2\pi r(r + h)$, where $\pi = 3.14$, find A when
   a $r = 7.3$, $h = 9.6$ **b** $r = 42$, $h = 78$.

72 Given that $v = \dfrac{2gh}{m}$, find v when
   a $g = 9.81$, $h = 5.7$, $m = 3.8$ **b** $g = 9.81$, $h = 154$, $m = 9.4$.

73 Given that $P = \dfrac{pv}{t} \times \dfrac{T}{V}$, find P when
   a $p = 12.6$, $v = 2.7$, $t = 307$, $T = 315$, $V = 4.5$
   b $p = 32.7$, $v = 8.95$, $t = 298$, $T = 301$, $V = 6.05$.

74 Given that $\dfrac{1}{R} = \dfrac{1}{r_1} + \dfrac{1}{r_2}$, find R when
   a $r_1 = 4$, $r_2 = 2$ **b** $r_1 = 4.15$, $r_2 = 2.36$.

Computation with numbers less than one

Introduction

Copy and complete these two tables.

Number	As a power of ten	Its logarithm
10 000	10^4	4
1000		
100		
10		
1		
0.1		
0.01		
0.001		
0.0001		

Number	Its logarithm
64 500	
6450	
645	
64.5	
6.45	
0.645	
0.0645	
0.00645	
0.000645	

Part 1 Multiplication

Work these.

1 3.07×0.184

2 1.27×0.516

3 2.53×0.361

4 0.127×4.82

5 6.72×0.328

6 5.55×0.417

7 0.805×4.96

8 0.212×6.35

9 $5.43 \times 7.24 \times 0.842$

10 $3.71 \times 5.63 \times 0.933$

11 $7.02 \times 0.662 \times 3.55$

12 $0.362 \times 6.74 \times 8.58$

13 5.64×0.0147

14 5.64×0.0633

15 0.0718×4.03

16 0.0205×8.75

17 23.2×0.232

18 0.624×14.6

19 195×0.195

20 0.347×214

21 4370×0.155

22 3870×0.0231

23 27.8×0.0311

24 50.6×0.0144

25 47.8×0.478

26 70.1×0.324

27 527×0.665

28 0.321×4390

29 81.7×0.0817

30 25.5×0.0637

31 0.00571×47.0

32 $246 \times 17.2 \times 0.000689$

33 0.391×0.247

34 0.183×0.472

35 0.00351×0.203

36 $0.0421 \times 0.135 \times 0.129$

37 0.624×0.539

38 0.367×0.544

39 0.885×0.0919

40 0.0437×0.593

41 $0.341 \times 0.942 \times 0.617$

42 $0.0475 \times 0.877 \times 0.55$

43 $0.723 \times 0.0512 \times 0.335$

44 $0.41 \times 0.607 \times 0.0748$

Part 2 Problems

Take $\pi = 3.14$ where necessary.

1 A rectangle has a length of 4.67 m and a width of 0.855 m. Find its area.

2 A bathroom mat is 2.46 m long and 0.76 m wide. What is its area?

3 A road is 66.5 km long and 0.0055 km wide. What area of tarmac (in km^2) is needed to cover the road?

4 A kitchen larder has a floor 0.85 m by 0.64 m. The larder is 2.36 m high. Find
a   the area of its floor in m^2 b   the volume of the larder in m^3.

5 A crate has a square cross-section of side 0.73 m and a length of 2.15 m. Find
a   the area of the cross-section in m^2 b   its volume in m^3.

6 Find the volume (in m^3) of a metal bar which is 12.9 m long with a rectangular cross-section 0.35 m by 0.14 m.

Computation with numbers less than one

7 A wooden stake is 3.06 m long with a rectangular section 0.17 m by 0.075 m. Find the volume of the stake in cubic metres.

8 Find the volume of
   a a sugar cube with edges 0.85 cm long
   b a stock cube with edges 0.94 cm long.

9 Find the volumes of these three pyramids. Give the answers in cubic metres.

a 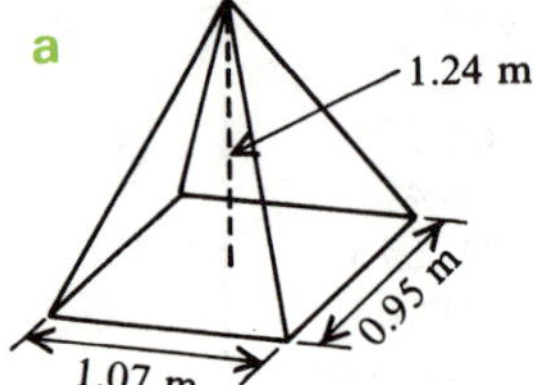

b 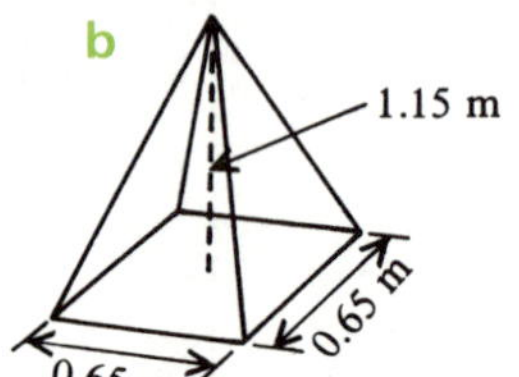

c 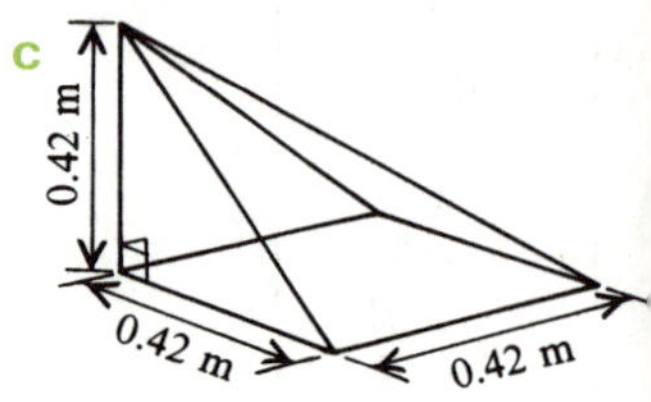

10 A ballpoint pen costs £0·37. Find the cost of 2370 pens (to the nearest £).

11 One kilogram of sugar costs £0·28. Find the cost of 3140 kg (to the nearest £).

12 1 km^2 of land is seeded with 0.84 tonnes of grass seed. What mass of seed is needed for 0.625 km^2 of land which is to be seeded in the same way?

13 If one litre of alcohol has a mass of 0.796 kg, what is the mass (in kg) of 23.5 litres?

14 If $1 = £0·56, what is
   a $0.50   b $0.73 in £ (to the nearest p)?

15 If 1 DM = £0·28, what is
   a 0.50 DM   b 0.65 DM (to the nearest p)?

16 Find the area (in m^2) of a circle with a radius of
   a 0.145 m   b 0.535 m   c 0.876 m.

17 Find   a the radius   b the area of a semicircle of diameter 1.43 metres.

18 A circle is drawn inside a square (of side 0.95 m) so that it touches each side of the square. Draw a diagram and find
   a the area of the square (in m^2)
   b the radius of the circle (in m)
   c the area of the circle (in m^2)
   d the area of the region which is inside the square but outside the circle.

19 Find the volume of a cylinder which has
   a a base radius of 0.65 m, a height of 2.25 m
   b a base diameter of 1.43 m, a height of 0.95 m.

20 A wooden cylinder has a height of 4.65 cm and a radius of 0.84 cm. Find
   a the area of the end of the cylinder
   b the volume of the cylinder
   c its mass, if 1 cm^3 of wood has a mass of 0.95 grams.

21 A broom handle is 1.87 m long and has a diameter of 0.035 m. Find
   a the radius of the handle   b the volume of the handle in m^3.

22 Find the volume of a cone with
   a a base radius of 0.58 m and a height of 3.34 m
   b a base diameter of 1.37 m and a height of 0.88 m.

23 A pile of sand has a conical shape of base diameter 4.7 m and height 0.76 m. Find
   a its base radius   b its volume   c its mass, if 1 m^3 has a mass of 1750 kg.

Computation with numbers less than one

24 Find the volume of a sphere of radius 0.86 m, giving the answer in m^3.
25 A mixing bowl has a hemispherical shape with a diameter of 0.37 m. Find
   a the radius of the bowl in metres   b its volume in m^3
   c its volume in litres, if 1 m^3 = 1000 litres
   d the mass of cooking oil which will fill the bowl, if one litre of oil has a mass
     of 0.82 kg.

Part 3 Division

1 $0.733 \div 3.32$
2 $0.885 \div 5.14$
3 $0.902 \div 2.61$
4 $0.558 \div 3.04$
5 $0.00717 \div 6.02$
6 $0.00413 \div 1.35$
7 $0.000954 \div 2.71$
8 $0.0000671 \div 5.23$
9 $0.925 \div 14.6$
10 $0.603 \div 27.4$
11 $0.881 \div 367$
12 $0.407 \div 156$
13 $0.0698 \div 34.2$
14 $0.00817 \div 55.6$
15 $0.0905 \div 607$
16 $0.000711 \div 320$

17 $24.8 \div 33.4$
18 $49.7 \div 65.2$
19 $60.2 \div 83.1$
20 $\dfrac{395}{506}$
21 $\dfrac{158}{551}$
22 $\dfrac{825}{896}$
23 $29.4 \div 482$
24 $706 \div 8540$
25 $15.4 \div 5040$
26 $\dfrac{78.6}{145}$

27 $\dfrac{98.5}{259}$
28 $\dfrac{67.3}{3450}$
29 $4.88 \div 7.96$
30 $2.44 \div 6.40$
31 $1.99 \div 4.31$
32 $\dfrac{2.94}{37.6}$
33 $\dfrac{6.72}{741}$
34 $\dfrac{9.07}{52.3}$

Part 4

1 $0.371 \div 5.62$
2 $0.454 \div 8.02$
3 $0.745 \div 9.12$
4 $0.00384 \div 6.50$
5 $0.00513 \div 7.78$
6 $0.000489 \div 6.22$
7 $0.497 \div 79.6$
8 $0.288 \div 30.1$
9 $0.601 \div 81.2$
10 $0.0489 \div 54.3$
11 $0.0294 \div 35.6$
12 $0.00147 \div 225$
13 $0.521 \div 476$
14 $0.883 \div 247$
15 $0.0914 \div 67.3$

16 $7.92 \div 0.327$
17 $6.91 \div 0.195$
18 $7.33 \div 0.552$
19 $\dfrac{4.46}{0.0164}$
20 $\dfrac{9.51}{0.00206}$
21 $\dfrac{7.72}{0.00641}$
22 $54.4 \div 0.183$
23 $329 \div 0.133$
24 $49.5 \div 0.0317$
25 $\dfrac{79.1}{0.911}$

26 $\dfrac{53.5}{0.662}$
27 $\dfrac{10.8}{0.221}$
28 $242 \div 0.351$
29 $688 \div 0.941$
30 $4370 \div 0.505$
31 $\dfrac{30.7}{0.0624}$
32 $\dfrac{14.6}{0.00338}$
33 $\dfrac{562}{0.0905}$

Computation with numbers less than one

Part 5

1. $0.497 \div 0.254$

2. $0.511 \div 0.201$

3. $0.0882 \div 0.0448$

4. $0.0723 \div 0.503$

5. $0.0243 \div 0.148$

6. $0.00668 \div 0.0414$

7. $0.945 \div 0.0314$

8. $0.789 \div 0.0511$

9. $0.0617 \div 0.00183$

10. $\dfrac{2.08}{0.415}$

11. $\dfrac{4.47}{0.546}$

12. $\dfrac{3.84}{0.0775}$

13. $0.287 \div 0.708$

14. $0.144 \div 0.342$

15. $0.0317 \div 0.0663$

16. $0.651 \div 0.0852$

17. $0.143 \div 0.0717$

18. $0.0435 \div 0.00622$

19. $\dfrac{0.635 \times 12.8}{37.4}$

20. $\dfrac{247 \times 0.845}{3240}$

21. $\dfrac{21.5}{3.27 \times 9.04}$

22. $\dfrac{0.932}{4.67 \times 2.43}$

23. $\dfrac{4.89}{72.3 \times 0.276}$

24. $\dfrac{15.3 \times 6.74}{8.42 \times 20.3}$

25. $\dfrac{0.845 \times 7.64}{16.3 \times 0.642}$

26. $\dfrac{0.707 \times 18.5}{0.315}$

27. $\dfrac{247 \times 0.046}{0.507}$

Part 6 Problems

1. A rectangle has an area of 0.925 m² and a length of 1.32 m. Find its width.

2. A coffee table top is 1.63 m long with an area of 0.89 m². Find its width.

3. a 1000 sheets of paper make a pile 110 mm high. How thick is one sheet?
   b 954 sheets of a different paper make a pile 763 mm high. How thick is one of these sheets?

4. A book of 540 pages is 6.75 cm thick.
   a How many sheets of paper are needed?
   b How thick is one of these sheets?

5. A 3.5-metre length of dress material contains 7150 cross-threads (or weft). How thick is one of these threads
   a in cm   b to the nearest tenth of a millimetre?

6. An air-conditioning unit circulates 15.5 m³ of air in 95 seconds. What is the rate at which air is flowing in m³ per minute?

7. Water from a tap fills a tub in 2 minutes 25 seconds. If the tub holds 95 litres, find the rate at which the water is flowing in litres per second.

8. If 15.7 cm³ of ice has a mass of 14.5 grams, find the mass of 1 cm³ of ice.

9. 0.518 cm³ of mercury has a mass of 6.94 grams. Find the density of mercury (i.e. the mass of 1 cm³).

10. If 0.054 cm³ of gold has a mass of 1.04 grams, find the density of gold in grams per cm³.

Computation with numbers less than one

11 If 24.10 francs = £2·45, how much is 1 franc worth in £'s?

12 If £0·65 = \$1.26, find how many dollars there are to £1.

13 If £0·84 = 1360 lire, find how many lire there are to £1.

14 A young boy piles 145 pennies on top of each other. If the height of the pile is 14.1 cm, what is the thickness of one penny to the nearest tenth of a mm?

15 A path is made from 136 paving slabs laid end to end. If the path is 102 metres long, what is the length in metres of one slab?

16 A car travels 150 miles in 3 hours 8 minutes. What is the average distance it travels each minute?

17 8.6 litres of petrol are fed into an engine in 37 minutes. What is the fuel consumption of the engine in litres per minute?

18 A cardboard box used for packing a washing machine has a volume of 0.92 m^3. Its length is 1.35 m and its width is 0.85 m. Find
  a the area of the base of the box
  b the height of the box.

19 A rectangular sheet of glass has a volume of 1590 cm^3. It is 95.5 cm long and 47.5 cm wide. Find
  a the area of the face of the glass   **b** the thickness of the glass.

20 A lorry loaded with 9.2 m^3 of turf delivers it to a garden where it is laid as a rectangular lawn 20.7 m by 7.4 m. Calculate
  a the area of the lawn
  b the thickness of the turf   (i) in metres   (ii) to the nearest cm.

21 A road 3.5 km long and 7.6 m wide is resurfaced with tarmac. The amount of tarmac required is 6820 m^3. Find
  a the surface area of the road in m^2
  b the thickness of the tarmac   (i) in metres   (ii) in centimetres.

22 The radius of a circle can be calculated from its circumference using the formula
$r = \dfrac{C}{2\pi}$. Calculate r, when
  a $C = 4.7$ cm, $\pi = 3.14$ **b** $C = 0.85$ cm, $\pi = 3.14$.

23 Use the formula $h = \dfrac{v^2}{2g}$ to find h, when
  a $g = 9.81$, $v = 2.33$ **b** $g = 9.81$, $v = 0.47$.

24 Given that $\alpha = \dfrac{e}{L\theta}$, find α when
  a $e = 2.45$, $L = 142$, $\theta = 16$ **b** $e = 0.76$, $L = 54$, $\theta = 12.5$.

25 Given that $R = \dfrac{r_1 r_2}{r_1 + r_2}$, find R when
  a $r_1 = 6.45$, $r_2 = 0.82$ **b** $r_1 = 0.97$, $r_2 = 0.66$
  c $r_1 = 0.46$, $r_2 = 0.35$.

Computation with numbers less than one

Part 7 Powers and roots

Work the following.

1	0.263^2	7	0.0215^3	13	0.517^4	19	0.0893^3
2	0.244^2	8	0.0152^4	14	0.604^4	20	0.0964^4
3	0.159^3	9	0.404^2	15	0.812^5	21	0.00162^2
4	0.183^3	10	0.514^2	16	0.915^6	22	0.00935^3
5	0.143^4	11	0.363^3	17	0.0334^2	23	0.95^{10}
6	0.0312^2	12	0.418^4	18	0.0591^2	24	0.8^{10}

25 Find the area of a square with side of length
   a   0.675 metres   b   0.038 km.

26 Find the volume of a cube with edges of length
   a   0.675 metres   b   0.038 km.

27 A cubical water tank is open at the top and its edges are 0.94 m long. Find
   a   the area of the base of the tank
   b   the area of sheet metal required to make the tank
   c   the volume of the tank
   d   the time taken to fill the tank from empty with water flowing in at a rate of
      0.85 cubic metres per minute.

28	$\sqrt{0.0735}$	36	$\sqrt[4]{0.000811}$	44	$\sqrt{0.651}$	52	$\sqrt{0.00355}$
29	$\sqrt{0.0487}$	37	$\sqrt[5]{0.0000602}$	45	$\sqrt[3]{0.584}$	53	$\sqrt{0.0082}$
30	$\sqrt{0.0923}$	38	$\sqrt[5]{0.000075}$	46	$\sqrt[3]{0.327}$	54	$\sqrt[4]{0.00373}$
31	$\sqrt{0.0794}$	39	$\sqrt{0.000557}$	47	$\sqrt[3]{0.650}$	55	$\sqrt[4]{0.0373}$
32	$\sqrt[3]{0.00862}$	40	$\sqrt{0.00067}$	48	$\sqrt[3]{0.0199}$	56	$\sqrt[4]{0.373}$
33	$\sqrt[3]{0.00467}$	41	$\sqrt{0.000009}$	49	$\sqrt[3]{0.0751}$	57	$\sqrt[5]{0.610}$
34	$\sqrt[3]{0.00874}$	42	$\sqrt{0.734}$	50	$\sqrt[3]{0.029}$		
35	$\sqrt[4]{0.000744}$	43	$\sqrt{0.557}$	51	$\sqrt{0.00173}$		

58 Find the length of side of a square whose area is
   a 0.084 m^2 b 0.84 m^2.

59 Find the length of edge of a cube whose volume is
   a 0.0065 m^3 b 0.65 m^3.

60 A packing-case has the shape of a cube with a volume of 0.93 m^3. Find
   a   the length of an edge of the case
   b   the area of one face of the case
   c   the total surface area of the case (to 2 significant figures).

Computation with numbers less than one

Part 8 A mixture

Work these using logarithms or a calculator.

1 2.67×0.362

2 0.845×4.98

3 0.246×0.663

4 $\dfrac{8.4}{12.5}$

5 $\dfrac{67.3}{325}$

6 $\dfrac{0.843}{2.07}$

7 $\dfrac{0.635}{8.19}$

8 $\dfrac{6.44 \times 1.85}{32.2}$

9 $\dfrac{0.88 \times 3.75}{13.6}$

10 $\dfrac{0.452 \times 0.836}{2.13}$

11 $\dfrac{0.672 \times 0.74}{5.53}$

12 $\dfrac{42.6}{0.845 \times 90.4}$

13 $\dfrac{6.88}{31.4 \times 0.52}$

14 $\dfrac{0.0843}{2.43 \times 1.22}$

15 $\dfrac{0.079}{0.681 \times 3.73}$

16 $\left(\dfrac{6.79}{9.83}\right)^2$

17 $(0.274 \times 4.3)^2$

18 $(21.5 \times 0.035)^3$

19 $\left(\dfrac{0.0977}{2.6}\right)^3$

20 $\left(\dfrac{0.0466 \times 245}{13.1}\right)^4$

21 $\left(\dfrac{72.4 \times 0.63}{128}\right)^2$

22 $\sqrt{\dfrac{4.88}{91.6}}$

23 $\sqrt{0.024 \times 2.15}$

24 $\sqrt{0.872 \times 0.115}$

25 $\sqrt[3]{\dfrac{5.43}{3640}}$

26 $\sqrt[3]{\dfrac{0.843}{752}}$

27 $\sqrt[3]{0.0372 \times 0.235}$

28 $\sqrt{0.97 \times 0.472}$

29 $\sqrt{\dfrac{36.5 \times 0.074}{4.32}}$

30 $\sqrt[4]{\dfrac{0.069}{8.6}}$

Arithmetic in Society

Wage slips

A typical wage slip could look like this.

STAFF NO.	WEEK NO.	PERIOD ENDING	TAX CODE	EMPLOYEE
BASIC PAY	OVERTIME	BONUS	OTHER ADDITIONS	TOTAL GROSS PAY
INCOME TAX	PENSION	NAT. INS.	OTHER DEDUCTIONS	TOTAL DEDUCTIONS
				NET PAY

The top row gives personal details; the middle row gives the different types of pay;
and the bottom row gives the various deductions from pay.

NET PAY = TOTAL GROSS PAY − TOTAL DEDUCTIONS.

1 Find the net pay for these four employees.

	Name	Total gross pay £	Total deductions £
a	A. Hawkins	98·45	21·34
b	F. Jewitt	82·45	20·65
c	D. Marple	71·52	15·27
d	G. Osbourne	124·35	36·18

2 Find the total deductions from the pay of these four employees.

	Name	Total gross pay £	Net pay £
a	M. Haworth	91·58	78·23
b	A. Kelly	89·34	72·55
c	S. Francis	108·26	87·72
d	P. Goodwin	145·08	106·36

3 Find the total deductions from the pay of these four employees.

	Name	Tax £	Pension fund £	Nat. Ins. £
a	C. Roberts	38·60	9·42	8·74
b	S. Brown	67·42	18·75	6·47
c	A. Davidson	104·56	36·41	41·26
d	P. Andrews	94·35	29·43	32·05

4 Find the total gross pay for each of these four employees.

	Name	Basic pay £	Overtime £	Bonus £	Back-pay £
a	J. James	83·74	27·68	15·42	10·74
b	E. Suter	75·60	12·47	8·42	24·81
c	B. Humble	120·42	6·79	12·85	8·49
d	D. Manns	98·67	20·42	18·70	9·78

Wage slips

5   These two wage slips have mistakes on them. Check the calculations and find the correct net pay in each case.

a

STAFF NO.	WEEK NO.	PERIOD ENDING	TAX CODE	EMPLOYEE
00124	02	16th April	–	J. AUSTIN

BASIC PAY	OVERTIME	BONUS	OTHER ADDITIONS	TOTAL GROSS PAY
105.74	12.34	–	–	118.08

INCOME TAX	PENSION	NAT. INS.	OTHER DEDUCTIONS	TOTAL DEDUCTIONS
21.42	12.46	7.47	6.42	42.15

				NET PAY
				75.93

b

STAFF NO.	WEEK NO.	PERIOD ENDING	TAX CODE	EMPLOYEE
01007	05	5th May	–	M. BERRY

BASIC PAY	OVERTIME	BONUS	OTHER ADDITIONS	TOTAL GROSS PAY
95.62	15.74	6.89	3.77	112.02

INCOME TAX	PENSION	NAT. INS.	OTHER DEDUCTIONS	TOTAL DEDUCTIONS
30.46	8.17	7.62	–	44.35

				NET PAY
				67.67

6   June Anderson works as a secretary. Copy the blank wage slip at the start of this section and fill in the relevant boxes with this information.

Her staff number is 4752, and week number 4 ends on 29/04/83.
Her basic pay is £52·75 and her overtime is £8·47.
She pays £11·24 in tax, £2·40 into a pension fund and £3·55 in National Insurance.

Calculate her total gross pay, her total deductions and her net pay.

7   Copy the blank wage slip again and fill in this information for Stephen Barnes who has a job as an apprentice engineer.

His staff number is 200493, and week number 12 ends on 25/06/83.
He earns £64·37 in basic pay, £6·21 in overtime and a bonus of £5·94.
He pays £19·47 in tax, £4·21 into a pension scheme and £4·42 in National Insurance.

Calculate his total gross pay, his total deductions and his net pay.

8   Calculate the net pay for these five employees.

		Basic pay £	Overtime £	Tax £	Pension £	Nat. Ins. £
a	JOHN	36·40	4·25	7·54	1·54	2·37
b	JANE	42·44	5·70	9·75	2·05	2·89
c	ERNEST	72·65	15·20	21·47	6·13	5·12
d	GEORGE	86·13	12·70	24·15	8·13	5·84
e	HARRY	96·47	21·13	30·41	11·20	7·49

9   Which one of these is equivalent to 'Superannuation'?

   a overtime   **b** bonus pay   **c** a pension fund   **d** National Insurance

10   Which two of these help finance the National Health Service, Sickness Benefits, Unemployment Benefits etc?

   a basic pay   **b** National Insurance   **c** overtime pay   **d** Income Tax

Overtime and rates of pay

1 **a** The working week is 42 hours at a basic rate of pay of £2·40 per hour. What is the basic weekly wage?

    b The working week is 36 hours at a basic rate of pay of £1·85 per hour. What is the basic weekly wage?

    c The working week is 38 hours at a basic rate of pay of £2·26 per hour. What is the basic weekly wage?

2 Find the basic weekly wage in each of these cases.

	No. of hours worked	Basic rate per hour
a	34	£2·36
b	38	£2·47
c	32	£3·25
d	36	£2·85
e	35	£3·14
f	40	£2·98

3 A man's basic rate is £2·60 per hour for a 42-hour week.
Find   **a** his basic weekly wage.

On Saturday he works 4 hours extra on 'double time'.
Find   **b** the hourly rate of pay on 'double time'
        c his overtime pay
        d his total wage for that week.

4 A woman's basic rate is £2·82 per hour for a 38-hour week.
Find   **a** her basic weekly wage.

One week she works an extra 5 hours on 'double time'.
Find   **b** the hourly rate of pay on 'double time'
        c her overtime pay
        d her total wage for that week.

5 Joe Saltby's working week is 40 hours long at a rate of pay of £3·15 per hour.
Find   **a** his basic weekly wage.

Last week he also worked 5 hours overtime on 'double time'.
Find   **b** the hourly rate of pay on 'double time'
        c his overtime pay
        d his total wage for the week.

6 Another workman's basic rate of pay is £2·40 per hour for a 36-hour week.
Find   **a** his basic weekly wage.

During one week, he works 4 hours overtime on 'time and a half'.
Find   **b** the overtime rate of pay per hour on 'time and a half'
        c his overtime pay
        d his total weekly wage for this particular week.

7 Mrs Fell works a basic 36-hour week earning £4·20 per hour.
Find   **a** her basic weekly wage.

Last week she worked another 4 hours on 'time and a half'.
Find   **b** the overtime rate of pay
        c her overtime pay
        d her total wage for the week.

Overtime and rates of pay

8 Another woman's basic rate of pay is £3·12 per hour for a 38-hour week.
Find   **a**   her basic weekly wage.

During a particular week, she puts in 6 hours on 'time and a half'.
Find   **b**   the overtime rate of pay per hour
         c   her overtime pay
         d   her total wage for this week.

9 A man's basic rate of pay is £3·10 for a 36-hour week.
Find   **a**   his basic weekly wage.

In one week he puts in 9 hours of overtime on 'time and a half'.
Find   **b**   the overtime rate of pay
         c   his overtime pay
         d   his total wage for this week.

10 For each person in this table find
   a   the basic weekly wage   **b**   the overtime rate of pay
   c   the overtime pay earned   **d**   the total weekly wage.

Name	Basic week hours	Basic rate per hour	Overtime		
			Time and a half (h)	Time and a quarter (h)	Double time (h)
Smith	42	£2·54	0	0	5
Jones	36	£2·90	4	0	0
Davis	38	£2·42	8	0	0
Mason	40	£2·40	0	5	0
Fell	42	£3·20	0	6	0

11 Mr Hanley applied for two jobs and was offered them both.
The first job paid £1·82 per hour for a 42-hour week.
The second job paid £2·14 per hour for a 40-hour week.
   a   Find the basic weekly wage for both these jobs.
   b   If the first job guarantees 6 hours of overtime every week on 'time and a half', what would he earn on overtime?
   c   If the second job has no overtime and Mr Hanley does not mind how long he spends at work, in which job would he earn more money and how much would this be?
   d   Which job would you choose, and why?

12 Mrs Bamforth has a choice of two jobs.
One is a part-time job for 26 hours a week paying £2·34 per hour, with no overtime.
The other is full-time for a 32-hour week paying £1·84 per hour, and there is the chance of 4 hours overtime on 'time and a half'.
   a   Find the weekly wage of the part-time job.
   b   Find the basic weekly wage of the full-time job.
   c   How much extra would she earn doing the 4 hours overtime?
   d   What is the total weekly wage of the full-time job, including the overtime?
   e   Which of these two jobs would you choose, and why?

13   **a**   You work a basic five-day week from 9 a.m. to 5 p.m. with one unpaid hour off for lunch. How many hours do you work in your basic week? If the basic rate is £3·68 per hour, what is your basic weekly wage?

Overtime and rates of pay

b This week you work on Saturday morning from 9 a.m. to 12 noon and earn 'time and a quarter'. What is the rate of pay on 'time and a quarter' and how much will this overtime pay be?

c What will be your total wage for the week?

14 Mrs Kayley works as a part-time secretary earning £3·40 per hour. She works all day on Monday and Thursday (with one unpaid hour off for lunch), mornings only on Tuesday and Friday, and not at all on a Wednesday.

The table shows the hours she works.

Mon.	9 a.m.–5 p.m.
Tues.	9 a.m.–12 noon
Wed.	—
Thurs.	9 a.m.–5 p.m.
Fri.	9 a.m.–12 noon

a How many hours per week does she work?

b What is her basic weekly wage?

c If she works Wednesday morning from 9 a.m. to 12 noon, she earns 'time and a quarter. How much extra will she then earn?

15 A firm employs Mr Greenwood as a part-time gardener. His basic working week is from 9 a.m. to 2 p.m. on Mondays and Wednesdays, and from 9 a.m. to 12 noon on Tuesdays, Thursdays and Fridays. He is paid for all his breaks, and his rate of pay is £2·80 per hour.

a How many hours does he work in one week?

b What is his basic weekly wage?

c During certain times of the year he works through to 4 p.m. on Tuesdays and for this overtime he is paid 'time and a quarter'. How much extra does he earn for such a Tuesday?

16 Mrs Firth should work from 9 a.m. to 5 p.m. from Monday to Friday with one unpaid hour off for lunch. How many hours should she work

a in a day   **b** in a week?

She has to clock-in and clock-out, and her time-card for last week is shown here. If she is over 10 minutes late in arriving or over 10 minutes early in leaving, she loses a quarter-hour's pay.

On which day did she lose pay for

c arriving late   **d** leaving early?

	In	Out
Mon.	9.02	5.01
Tues.	9.04	5.00
Wed.	8.57	4.59
Thurs.	9.12	5.02
Fri.	8.59	4.48

e How many hours of work was she paid for during this week?

f If her hourly rate of pay is £3·32, what were her earnings for this week?

17 A bus conductor's times of duty for a particular week are shown in this table. He works Saturday but has Tuesday free. His basic rate of pay is £5·60 per hour, but on Saturday he earns 'time and a quarter'. If he works a 'split shift', where he has a break in his duty of over three hours, then he earns an extra £4 for that duty. Calculate

Mon.	5.30 a.m.–9.30 a.m. and 3 p.m.–7 p.m.
Tues.	—
Wed.	6 a.m.–2 p.m.
Thurs.	7 a.m.–10 a.m. and 4.30 p.m.–10 p.m.
Fri.	1 p.m.–8.30 p.m.
Sat.	4 p.m.–midnight

a how many hours he works from Monday to Friday and his total earnings for these days including any extra for 'split shifts'

b his rate of pay for Saturday and his earnings on Saturday

c his total earnings for the week.

Overtime and rates of pay

18 From Monday to Friday Mr Jackson works from 8 a.m. to 5 p.m. with one unpaid hour off for lunch. How many hours should he work
    a in a day   **b** in a working week of five days?

However, he has to clock-in and clock-out and his time-card for a particular week is shown here.

If he is more than 10 minutes late, he loses a quarter-hour's pay.

If he leaves more than 10 minutes early, he also loses a quarter-hour's pay.

	In	Out
Mon.	8.02	5.03
Tues.	7.59	4.59
Wed.	8.14	5.01
Thurs.	8.13	4.45
Fri.	8.01	4.46
Sat.	8.03	12.02

  c Look at the time-card to find on which three days of the week he lost pay.
  d How many quarter-hour's pay did he lose?
  e How many hours in the week did he get paid for (excluding Saturday)?
  f If his basic rate of pay is £2·65 per hour, what was his basic pay for this week (excluding Saturday)?
  g On Saturday he earns 'double time'. How much overtime pay did he earn on this Saturday?
  h What were his total earnings for the week?

Piece-work

1 Mrs Jameson works part-time in a dry-cleaners where she gets paid 8 pence for every shirt she irons.
   a If on Monday she irons 115 shirts, how much is she paid?
   b If on Tuesday she irons 128 shirts, how much is she paid?
   c If on Wednesday she irons 134 shirts, how much is she paid?
   d She does not work on Thursday or Friday, so what are her total earnings for the week?

2 Penny Mallins does some typing at home. She charges 45 pence per page. One week she types a report for a local firm.
   On Monday she types 25 pages.   On Tuesday she types 38 pages.
   On Wednesday she types 42 pages.   On Thursday she types 27 pages.
   On Friday she types 14 pages.
   a How many pages did she type altogether?
   b What did she charge the firm for the whole report?

3 Two school children pick peas during three days of their holiday. They get paid 35 pence for every crate they fill. The number of crates each fills is shown in this table.

	Mon.	Tues.	Wed.
Louise	9	11	14
Stephen	7	12	12

   a How many crates does Louise fill altogether?
   b How much does Louise earn during the three days?
   c How many crates does Stephen fill altogether?
   d How much does Stephen earn during the three days?

4 A mail-order firm pays 2p for every envelope which is filled, sealed and addressed. Four girls deal with the given number of envelopes in one day. Find how much each girl earns and give the answers in £.

Name	Ann	Jo	Sue	Nina
No. of envelopes	243	324	186	255

5 Homeknit Ltd employ people at home to knit clothes. They provide the wool and pay £3·50 for each sleeved jumper, £2·90 for each sleeveless jumper, £1·05 for each pair of socks and £1·35 for each woollen cap.
   Four people make the following articles in one week. Find how much each earns.

		Sleeved jumpers	Sleeveless jumpers	Pairs of socks	Caps
a	Mrs Fuller	2	4	0	5
b	Mrs Warren	6	0	4	0
c	Mr Hopkins	3	4	6	0
d	Mrs Cook	4	2	5	3

6 In his spare-time, Jim Calder paints houses. He charges £2·60 for each small window, £4·25 for each large window, £4·10 for each door and 35 pence for each metre of guttering.
   One weekend he paints two houses which require these items painting.

	Small windows	Large windows	Doors	Metres of guttering
First house	3	4	2	21
Second house	5	2	2	14

How much does he charge for each house?

Piece-work

7 One weekend, Alan Blacker works for his uncle in his allotment. He is paid 8 pence for each square metre of ground which he digs over, 4 pence for each row of vegetables which he weeds and 2 pence for each kilogram of potatoes which he lifts.

If Alan digs over a rectangular plot 4 metres by 12 metres, weeds 13 rows of vegetables and lifts 34 kg of potatoes, how much does his uncle have to pay him?

8 On Monday to Saturday a newsagent pays his paperboys 2p for each newspaper or magazine which they deliver and on a Sunday he pays double this amount.

Lucy Hanman delivers 124 papers on her round each day from Monday to Saturday. The number of magazines she delivers each day is shown in this table.

	Mon.	Tues.	Wed.	Thurs.	Fri.	Sat.
No. of magazines	25	14	32	21	46	12

She also has a Sunday round of 205 papers with 62 magazine supplements.
  a How many magazines does she deliver from Monday to Saturday?
  b How much does she earn from Monday to Saturday?
  c What are her total earnings for the week?

9 Sandra Mallinson enters a 42-mile sponsored walk for the Save The Children Fund. Her mother sponsors her at 8p per mile, her Great-aunt Mary sponsors her at 12p per mile, and her father said he would give her 5p per mile for the first half of the walk and 15p per mile for the second half of the walk.
  a If she completes the walk, how much will she collect?
  b If she walks only 35 miles, how much will she collect?

10 Mrs McDonald weaves Harris tweed in two widths. She sells the narrow width at £3·25 per metre length and the wider width at £3·65 per metre length. It takes her 30 minutes to weave one metre of the narrow cloth and 40 minutes to weave one metre of the wider cloth.

She works for 6 hours each working day, spending Monday and Tuesday on the narrow tweed and Wednesday, Thursday and Friday on the wider tweed. On Saturdays and Sundays she does no weaving.
  a How many metres of each type of cloth does she weave in one week?
  b How much will she get from selling one week's output of tweed?

11 John Powers is a student who earns £8·45 a day crating carpets as a holiday job. If he crates more than 50 carpets a day, he gets an extra 25 pence for every carpet over 50.

Today he crated 62 carpets.
  a For how many carpets will he be paid extra?
  b How much extra will he be paid?
  c What did he earn altogether today?

12 A workman is paid a basic £12·84 a day to load ball-bearings into a mounting. If he loads more than 650 mountings in a day, he is paid £0·18 for every one over 650.
  a On Monday he loaded 684. For how many mountings was he paid extra and what extra amount did he receive?
  b What were his total earnings for this Monday?
  c On Tuesday he loaded 617 mountings. How many extra was this?
  d What were his total earnings for this Tuesday?

Piece-work

13 Mr Rodgers assembles electrical gadgets and is paid a basic rate of £2·14 per hour for an 8-hour day.
  a What are his basic earnings for one full day's work?
  b If he assembles more than 65 gadgets in a day, he is paid an extra £0·74 for each one over 65. Yesterday he assembled 72 gadgets, so how much extra did he earn?
  c What were his total earnings yesterday?

14 A window-cleaning firm employs Mr Kaye at a rate of £1·85 an hour for a 7-hour day.
  a What are his basic daily earnings for the full 7 hours?

He earns an extra 15 pence for each house over 45 that he visits in one day.
  b If today he visited 62 houses, how much extra did he earn?
  c What were his total daily earnings?

15 Mr Ellis makes metal components for a firm. He earns a basic weekly wage of £72·50, but he is paid an extra 42 pence for each component he makes over 450 in one day.

This table gives the numbers he made in the five days of a week.

	Mon.	Tues.	Wed.	Thurs.	Fri.
No. of Parts made	427	438	407	462	471

  a On which days did he make more than 450 components?
  b How much extra did he earn on each of those days?
  c What was his total weekly wage for this particular week?

16 Mrs Calder goes on a 'Sponsored Slim' for charity. She will get £18·50 if she loses 8 kg, and for every kilogram over 8 kg she will get £1·76.

Find how much she receives, if she loses
  a 7 kg   **b** 9 kg   **c** 14 kg.

17 A factory worker is paid a basic £2·13 per hour for an 8-hour day. If his machine produces more than 2500 plastic units in a day, he is paid an extra 2p for each unit over 2500.
  a What are his basic earnings per day?
  b How much extra would he earn on a day when he produced 2746 plastic units?
  c What would be his total pay on that day?

18 Mrs Crabtree works for a firm where she assembles children's toys. She is paid 14 pence for each one she assembles, and, if she assembles more than 800 in a working week from Monday to Friday, she can volunteer to work Saturday morning when she will be paid 18 pence for each toy.

During one week she assembles these numbers.

	Mon.	Tues.	Wed.	Thurs.	Fri.
No. of toys	185	121	156	174	172

  a How many toys did she assemble on these five days, and how much did she earn?
  b Did she qualify for Saturday working? If she assembles 86 toys on Saturday morning, what will her total pay for the week be?

Commission

1 A man sells insurance from door-to-door. He is paid a basic £62·50 a week.
   He also earns 8% commission on all insurance which he sells.
   Last week he sold £245 worth of insurance.
   a   Find 8% of £254.
   b   Find his total earnings for last week.

2 A salesman in a showroom earns a basic £42·75 a week, and he also gets
   2% commission on all the sales he makes.
   Last week he sold £3250 of goods.
   a   Find 2% of £3250.
   b   Find his total earnings last week.

3 A travel agent earned a basic £58·70 a week and then received a commission of
   4% on the £1765 worth of holidays she sold.
   a   Find the commission she made.
   b   What were her total earnings for the week?

4 A man earns a basic £55·40 a week taking photographs of holiday-makers at a
   seaside resort. He is then given a commission of 4% of the income from the
   photographs which people buy.
   a   If the income from the sale of photos is £248 in a week, what is his
       commission and what is his total pay for the week?
   b   If the income from the sales is £135, what is his commission and what is his
       total pay for the week?

5 An ice-cream salesman earns a basic £42·75 a week and also gets a 6%
   commission on his sales.
   a   If his weekly sales are £485, find his commission and also his total pay for
       that week.
   b   If his weekly sales are £207, find his commission and also his total pay for
       that week.

6 A man is paid £1·55 an hour to stand behind a market stall selling fruit and
   vegetables for 6 hours on a Saturday. He also gets 3% of the takings.
   a   Find his commission, if the takings are £284.
   b   Find his total earnings for the day.

7 Mrs Muskie earns a basic £1·47 an hour selling cosmetics from door-to-door.
   She also gets a commission of 12% on all her sales.
   a   If she works for 6 hours one day, what is her basic pay for the day?
   b   If in this time she sells £157 of cosmetics, what is her commission?
   c   How much are her total earnings for that day?

8 Frank Furtough is a hot-dog seller earning a basic £1·65 an hour. He is also
   paid a commission of 14% on all his sales.
   a   One evening he works for 5 hours in the city centre. What is his basic pay?
   b   Find his commission, if his takings are £65.
   c   What are his total earnings for the evening's work?

9 A double-glazing salesman earns a basic £2·38 an hour for a 42-hour week.
   In addition he gets a 12% commission on all the double glazing he sells.
   a   What is his basic weekly wage?
   b   If he sells £1428 of double glazing in a certain week, what commission does
       he earn?
   c   What is his total income for that week?

Commission

10 A newspaper seller stands outside the railway station for four hours each day and is paid a basic £1·45 per hour. He also gets 2% commission on his sales.

    a What is his basic pay each day?

    b What is his basic pay for a working week of 5 days?

    c If he sells £237 worth of newspapers during the week, what is his commission?

    d What are his total weekly earnings?

11 The salespeople employed by a firm have their monthly salaries £S calculated as shown in this flow chart.

Each salary comprises a basic salary £B to which is added a commission £C. Both £B and £C depend on the amount of sales made £X.

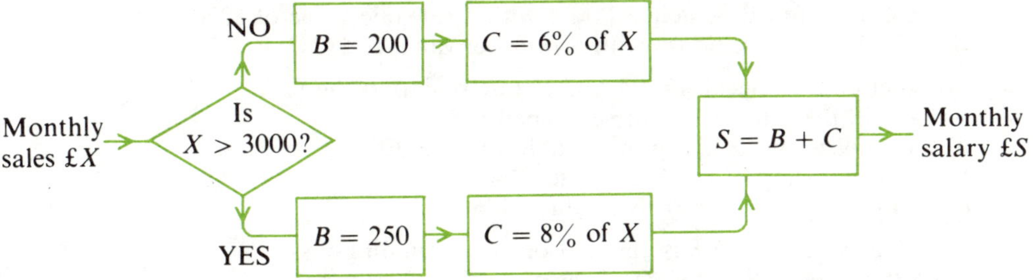

Calculate the salaries of the salespeople whose monthly sales £X are

    a £2500   **b** £3200   **c** £800   **d** £3850.

12 Another company calculates its salespeople's salaries as in this flow diagram.

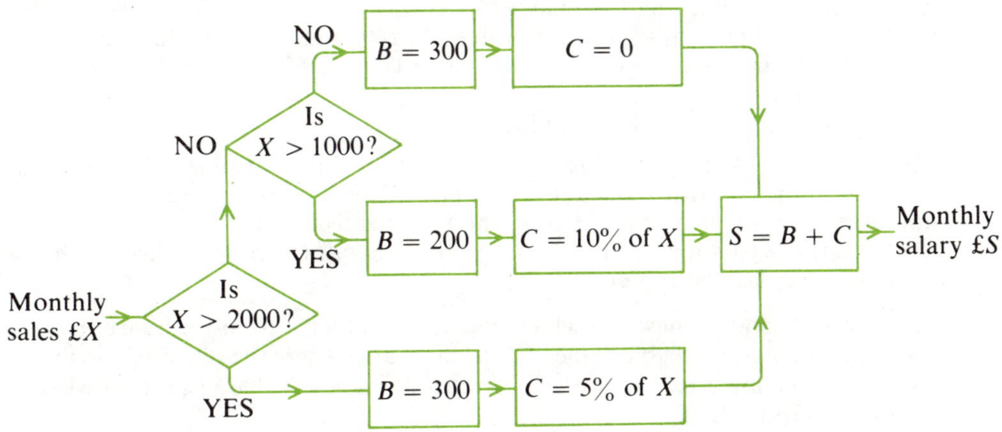

Calculate the salaries of the salespeople whose monthly sales are

    a £800   **b** £1400   **c** £1900   **d** £2600

    e £2000   **f** £1000.

Rates

A **rate** is a tax on buildings which is paid to a local authority or council so that it can provide services such as schools, police, libraries, sport centres, street lighting etc.

The **rateable value** of a property is assessed by a valuation officer.
(Do not confuse it with the price of the property, if it is sold.)

Each year the council decides what tax to impose for each £1 of rateable value – this tax is the rate in the £.

Part 1

1 A council decides that the rate will be 65p in the £.
   Find the rates paid on these properties.
   a a bungalow with a rateable value of £140
   b a 3-bedroomed semi-detached house with a rateable value £230
   c a 4-bedroomed detached house with a rateable value of £270
   d a terraced town house with a rateable value of £124

2 Another council decides on a higher rate of 84p in the £.
   What will the rates be on these properties?
   a a small terraced house of rateable value £130
   b a corner shop of rateable value £240
   c an office-block of rateable value £746

3 A rate of £1·24 in the £ is charged on these buildings.
   Find the rates paid for each of them.
   a a semi-detached house with a rateable value of £146
   b a farmhouse with a rateable value of £234
   c a supermarket with a rateable value of £365

4 Mr and Mrs Goulding move house but stay in the same local authority area, where a rate of 72p in the £ is charged.
   a If their old house has a rateable value of £160, find the rates they paid.
   b Their new house is larger and its rateable value is £215. Find the new rates which they pay?
   c By how much have their rates increased?

5 Mr Weary retired and moved from a large house with a rateable value of £265 to a smaller house with a rateable value of £172. If he stays in the same local authority area where a rate of 94p in the £ is charged, find
   a the rates he paid for his large house   b   the rates he pays for his new house
   c how much less rates he pays.

6 Jack Banks has got a new job which means he will have to move house. His present house has a rateable value of £164 in an area where the rate is 92p in the £. He intends buying a new house with a rateable value of £206 in an area where the rate is 73p in the £.
   a How much does he pay in rates at present?
   b How much will he pay in rates after the move?
   c Will he be spending more or less on rates, and by how much?

7 Woldby Local Authority charged a rate of 86p in the £ last year.
   They need to raise more money this year, and so they have increased the rate by 8p to 94p in the £.
   a If Mr and Mrs Paturrson live in a house with a rateable value of £162, how much did they pay in rates last year?
   b How much will their rates be this year?
   c What is the increase in their rates?

Rates

8 Grubthorpe Local Authority charge a rate of 71p in the £. Next year this rate will rise by 4p to 75p in the £.
   a If Mr Ali's house has a rateable value of £206, how much were his rates this year?
   b How much will he pay next year?
   c What will be the increase in his rates?

9 Your house has a rateable value of £184. A factory is built across the road and the smoke and noise from it are such a nuisance that the local authority reduce your rateable value to £155.
   a If the rate is £1·12 in the £, how much were your rates before the factory was built?
   b How much are your rates after the factory is built?
   c How much less do you now pay?

10 Mr Kinder and his family move into a house of rateable value £225. They have central heating and a shower-room installed. This increases their rateable value to £242. The council charges a rate of £1·26 in the £.
   a What rates did they pay before these improvements were made?
   b What rates are paid after the improvements have been made?
   c By how much did their rates increase because of the improvements?

11 Mrs Firth's house had a rateable value of £142 until she had a garage built next to it when the rateable value increased to £168.

   If the rate this year is 94p in the £, find
   a the rates she paid before the garage was built
   b the rates she paid after the garage was built
   c the increase in her rates.

12 A shop has a rateable value of £125 in an area where the rate is 84p in the £. The owner of the shop lives in a house with a rateable value of £248 in an area where the rate is 96p in the £.

   Find the total he pays in rates on his house and his shop.

Part 2

1 This table shows the rateable value of all the property in six local authorities. It also gives the amount of money they have to raise to pay for the services they give.

   Calculate the rate in the £ for each authority.

	Authority	Total rateable value of all property (£)	Amount to be raised by the rates (£)
a	Kirkly	2 000 000	1 600 000
b	Thornham	6 000 000	3 300 000
c	Moorton	8 750 000	7 000 000
d	Scarby	12 500 000	9 000 000
e	Grancastle	640 000	480 000
f	Rincaster	7 500 000	6 150 000

Rates

2 The rateable value of all the property in an area totals £2 500 000. If the council wants to raise £3 600 000, what rate in the £ should it levy?

3 Another authority has an area in which the total rateable value is £450 000. If it wants to raise £540 000, what rate in the £ should it charge?

4 A council wants to raise £1 400 000 to run its services. If the rateable value in its area totals £1 120 000, what rate in the £ should it charge?

5 A local council decides to spend these amounts of money on various services.

	Expenditure (£)
Education	15 525 000
Welfare	3 254 000
Police	2 557 000
Lighting	829 000
Roads, footpaths etc.	4 672 000
Libraries, museums etc.	682 000
Housing	1 027 000
Others	3 879 000

a Find the total amount to be spent on all these services.

b If the Government gives the council a grant of £8 425 000, how much is left for it to raise through the rates?

c If the rateable value of all property is £32 000 000, what must the rate in the £ be?

6 Another council wants to spend these amounts on services.

	Expenditure (£)
Education	3 251 000
Police	862 000
Welfare and Health	1 108 000
Roads, lighting etc.	1 015 000
Housing	268 000
Cleansing	379 000
Libraries, museums, parks etc.	726 000
Others	1 204 000

a What is the total amount the council wants to spend?

b How much of this will it have to raise through the rates, if central Government gives the council a grant of £1 013 000?

c If the rateable value of all property in its area is £7 400 000, what will the rate in the £ have to be?

7 The total amount a council wants to spend is £23 562 500 and central Government will give a grant of £9 062 500. If the total rateable value of the area is £12 500 000, find what rate in the £ the council will have to charge.

Rates

8 Central Government gives a grant of £7 552 000 towards the £17 152 000 which a council needs to spend on its services. What rate per £ will the council charge if the total rateable value of its area is £7 500 000?

9 This year central Government provided a grant of £2 175 000 towards the £8 475 000 which a council needed to spend on its services.
   a How much did the council have to raise by rates?

   b If the rateable value of the area is £4 500 000, what rate per £ is charged?

   c Next year, the council want to spend the same amount of money, but central Government will reduce its grant to £1 545 000. How much will the council have to raise from the rates and by how many pence will it have to increase the rate per £?

10 If Thornwick Council increases its rate by one penny, the yield will be £65 350. How much more does the council raise, if it increases the rate by
   a 4p   b 13p?

11 Each penny rate which Gainsby Council levies raises £38 450. How much is raised, if the rate increases by
   a 5p   b 14p?

12 Skemsworth Council needs to raise an extra £324 000. If each penny rate levied yields £81 000, by how much does it need to raise its rate per £?

13 A council needs to raise £281 250, and a penny rate will yield £62 500. What will it have to increase the rate per £ by to raise the required amount?

14 If a penny rate raises £94 750, how much will a rate of 126p in the £ raise towards the cost of services provided by a council?

15 A council charges a rate of 145p in the £. If a penny rate raises £72 540, what is the council's income from the full rate?

Income tax

Income tax is based on how much a person earns in one year. Not all of a person's income is taxed; the amounts not taxed are called **allowances**.

The **taxable income** is the gross income less the allowances.

The **tax paid** is a certain percentage of the taxable income. There is a **basic rate** of tax; but higher income earners will pay tax at a higher rate.

These calculations are listed in this flow diagram.

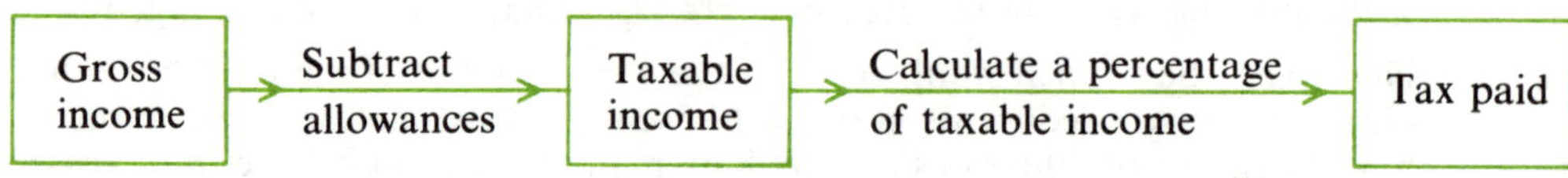

The Government can change the allowances and the rates of tax in the annual Budget. In these problems, use the values given here.

Allowances	£	Rates of tax		£
Single person	1785	Basic rate	30%	1–14 600
Married man	2795	Higher rates	40%	next 3000
Wife's earned income	1785		45%	next 4000
Dependent relative	100			… and others
… and others				

Part 1

1   Sandra Hyde is single and earns £3825 a year as an office typist. Find
   a her taxable income   **b** the amount of tax she pays in a year
   c her net income after tax has been paid.

2   Mr Hudson is married and his wife has no income. He earns £4225 a year as a lorry driver. Find
   a his taxable income   **b** the amount of tax he pays in the year
   c his net income after paying tax.

3   Mrs Spalding is a widow earning £1855 a year from a part-time job. Find
   a her taxable income   **b** the amount of tax she pays
   c her net annual income after tax has been paid
   d her net monthly income after tax, to the nearest penny.

4   Mr Withens is married and his wife has no earnings. In one year he earns £6550 as a fitter. Find
   a his taxable income   **b** the amount of tax he pays in one year
   c his net income after paying tax for the year
   d his net monthly income after tax, to the nearest penny.

5   Miss Shaw earns £5500 in one year and she looks after her mother at home. Find
   a her total allowances   **b** her taxable income
   c the amount of tax she pays in the year
   d her income after paying tax.

6   Mr and Mrs Delph look after both of Mrs Delph's parents. Mr Delph earns £7240 p.a. as a draughtsman. Calculate
   a his total allowances against tax   **b** his taxable income
   c the amount of tax he pays in a year.

Income tax

7 Mr Buckden earns £380 per month as a general labourer. He is married and his wife does not work. Find
   a   his gross annual income   b   his taxable income
   c   the amount of tax he pays during the year
   d   his monthly income after tax to the nearest £.

8 Mr Thorpe earns £335 per month in a factory and is not married. Find
   a   his gross annual income   b   his taxable income
   c   the amount of tax he pays in the year
   d   his annual income after tax had been deducted
   e   his monthly income after tax, to the nearest £.

9 Herbert Winchester is married and his wife looks after his elderly mother. If he earns £415 per month, find
   a   his gross annual income   b   his taxable income
   c   the amount of tax he pays during the year
   d   his monthly income after tax, to the nearest £.

10 Sally Bywell who is not married earns £43 a week as a hairdresser's assistant.
   a   How much does she earn in a year of 52 weeks?
   b   Find the amount of tax she pays in the year.
   c   Find her net yearly income after paying tax.
   d   What is her net weekly income after tax, to the nearest penny?

11 John Boothroyd earns £67 a week as a trainee engineer. He is not married.
   a   How much does he earn in a year of 52 weeks?
   b   Find the amount of tax he pays in the year.
   c   Find his net yearly income after paying tax.
   d   What is his net weekly income after tax, to the nearest penny?

12 Joanne Hickman is not married and has just started work as a trainee computer operator earning £76 per week. Find
   a   her yearly income over 52 weeks   b   her taxable income
   c   the amount of tax she pays during the year
   d   her weekly income after tax, to the nearest penny.

Part 2

1 A young unmarried girl earns £165 per month in her first job. Find
   a   her gross annual income   b   her taxable income
   c   the amount of tax she pays in the year
   d   her annual income net of tax
   e   her monthly income net of tax, to the nearest £.

2 Barry earns a basic £58 per week and also in a full year he earns an extra £544 overtime. If he is not married, find
   a   his total annual income   b   his taxable income
   c   his yearly payment of tax   d   his annual income after tax
   e   his monthly income after tax, to the nearest £.

3 John Gordon recently got married and his wife gave up her job. He earns £524 per month and gets a £342 bonus at the end of the year. Find
   a   his gross annual income   b   the amount of tax he pays for the year
   c   his income for the year after tax has been deducted
   d   his monthly income after tax, to the nearest £.

Income tax

4   Mr and Mrs Pearson both work. He earns a basic £4585 with overtime of £670 during the year and she earns £4720 for the year. Calculate
   a   their total annual income   b   their joint taxable income
   c   the total amount of tax they pay   d   their joint annual income net of tax.

5   Mr and Mrs Minchin both work as sales assistants. She earns £3870 p.a. and he earns £4250 p.a. Mrs Minchin gets a Christmas bonus of £255, Mr Minchin earns another £465 in overtime. Calculate
   a   their total annual income   b   their joint taxable income
   c   the total amount of tax they pay   d   their joint annual income net of tax.

6   Gordon McAndrew earns £7255 p.a. as an engineer and his wife has a part-time job earning £1875 p.a. Calculate
   a   their total annual income   b   their joint taxable income
   c   the total amount of tax they pay   d   their joint annual income net of tax.

7   Mrs Smithson earns £2·25 per hour working part-time for six hours each day on three days of the week for 36 weeks of the year. Her husband earns £7255 p.a. as a teacher. Find
   a   how much Mrs Smithson earns in the year
   b   the total income of Mr and Mrs Smithson   c   their taxable income
   d   the amount of tax they pay during the year
   e   their joint annual income net of tax.

8   Don Meltham earns a basic £4750 p.a. He also works an average of five hours overtime at £3·50 per hour for 35 weeks of the year. His wife has a job where she earns £3175 p.a. Find
   a   Don's total annual income   b   their joint total annual income
   c   the amount of tax paid
   d   their joint annual income after tax has been deducted.

9   The chief accountant of a large firm earns £18 545 p.a. He is married and his wife has no income. Calculate
   a   how much of his income will be taxed at a higher rate
   b   how much tax he pays at the basic rate
   c   how much tax he pays at the higher rate
   d   his annual income after tax has been paid.

10   Ernest Sykes has an income of £16 525 p.a. as the general manager of a firm. He is not married. Find
   a   how much of his income will be taxed at a higher rate
   b   how much tax he pays at the basic rate
   c   how much tax he pays at the higher rate
   d   his annual income after tax has been deducted.

11   Miss Walton is a headmistress earning £17 225 p.a. Calculate
   a   the total amount of tax she pays, both at the basic rate and at a higher rate
   b   her annual income after tax has been paid
   c   her net monthly income, to the nearest penny.

12   The director of a firm earns £19 895 p.a. He is a married man with a non-working wife. Find
   a   how much of his income will be taxed at a higher rate
   b   how much tax he pays at the basic rate
   c   how much tax he pays at the higher rate
   d   his annual income after tax has been paid.

Mortgages

1 Mr and Mrs Wilkinson borrowed £12 500 from a Building Society to buy their house. They have to pay 9% p.a. interest on this loan.
   a How much interest do they pay in the first year?
   b If they pay equal instalments each month, how much interest do they pay each month?

2 A young couple borrow £16 050 from a Building Society to buy a house, and they have to pay 8% interest on the loan each year.
   a How much interest do they pay in the first year?
   b How much interest do they pay each month of this first year?

3 You borrow £9500 from a Building Society which charges 12% interest each year.
   a What interest do you pay in the first year?
   b How much interest do you pay each month of the first year?

4 A bank charges interest of 13% p.a. on a house loan of £14 820.
   a How much interest is paid in one year?   b What is this per month?

5 Interest of 11% p.a. is charged on a mortgage of £8700. How much is this per month if there are twelve equal payments during the year?

6 Find the monthly repayments of interest during the first year on a loan of £9750 at 14% p.a.

7 Find the monthly repayments of interest during the first year on a mortgage of £16 540 loaned at 9% p.a.

8 What will be the interest payment each month of the first year on a loan of £18 660, if interest is reckoned at 13% p.a.?

Repayment mortgages

With this type of mortgage each monthly repayment is used partly to pay the interest on the loan and partly to reduce the size of the loan.

9 a Find the interest paid at 8% p.a. in one year on a loan of £12 600.
   b With monthly repayments of £93, how much is repaid in one year?
   c Copy and complete the table alongside to find how much of the loan is still outstanding after one year.

	£
Original loan + Interest	__________
− Repayments	__________
Loan outstanding	__________

10 a Find the interest paid at 7% p.a. in the first year on a loan of £16 800.
   b If monthly repayments are £106, how much is repaid in one year?
   c Copy and complete this table to find how much is still outstanding after one year.

	£
Original loan + Interest	__________
− Repayments	__________
Loan outstanding	__________

11 a How much interest at 13% p.a. is paid in one year on a mortgage of £14 400?
   b How much do the repayments total over the year, if each month they are £181?
   c How much of the loan is still outstanding after one year?

Mortgages

12 a What is the interest to be paid in the first year on a mortgage of £9480 taken out at 11% p.a.?

 b What are the total repayments in this year, if they are £97 each month?

 c To what amount has the mortgage been reduced by the end of the first year?

13 a How much interest is paid in the first year on a mortgage of £12 650 at 14% p.a.?

 b If monthly repayments are £172, what do they total in a year?

 c How much of the loan is outstanding at the end of the first year?

14 a A young couple take out a mortgage which requires monthly repayments of £75 during the whole period of the mortgage. How much do they pay each year?

 b The original loan was for £12 000 with interest charged at 5% p.a. How much interest is due during the first year?

 c Copy and complete part of this table to find how much is still outstanding after one year.

 d How much interest is due during the second year?

 e Complete part of the table to find how much is still outstanding after two years.

 f How much interest is due during the third year?

 g Finish the table to find the amount outstanding after three years.

	£
Original loan + Interest in 1st year	__________
− Repayments	__________
Loan after 1 year + Interest in 2nd year	__________
− Repayments	__________
Loan after 2 years + Interest in 3rd year	__________
− Repayments	__________
Loan after 3 years	__________

15

 a Mr and Mrs Dictas are given a mortgage requiring repayments of £50 each month of the loan. How much do they repay each year?

 b The original mortgage was for £10 000 with interest reckoned at 5% p.a. What interest do they pay during the first year?

 c Set out the work as in question **14** to find how much of the loan is still outstanding at the end of the first year. The flow diagram above will help.

 d Continue the working to show what they still owe after two years, and then after three years.

In problems **16–20**, round off each amount to the nearest £. (Calculations can be checked using a calculator.)

16 These figures describe a mortgage which is given to a family by a Building Society.

Original loan	£8000
Interest rate	10% p.a.
monthly repayments	£90

Set the working out as in question **14**, and carry on the calculations until the loan is reduced to nought. For how many years has the mortgage run?

Mortgages

17 Another mortgage is given with an original loan of £9500
an interest rate of 10% p.a.
and monthly repayments of £105.
Set the working out as before to find out for how many years the mortgage runs before it is paid off.

18 Set the working out as before to find how long it takes for a £12 000 mortgage to be paid off with an interest rate of 10% p.a. and monthly repayments of £110.

19 How long does it take to pay off a £14 500 mortgage with interest at 10% p.a. and monthly repayments of £135?

20 If the £14 500 mortgage of question **19** had interest charged at 5% and the monthly repayments stayed at £135, how many years would it take to pay off the debt?

21 For each of questions **16–20**, plot a graph of the outstanding loan against the number of years paid.
For each graph you draw, find how long it takes to pay off half the original mortgage.

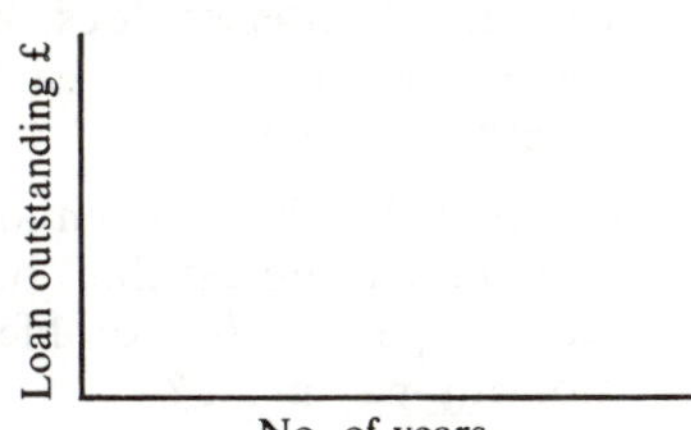

Endowment mortgages

With this type of mortgage, only the interest is repaid month by month. The original loan is not reduced at all, but is covered by a life insurance policy. When this policy ends, the mortgage is paid off all in one payment.

22 An endowment mortgage of £15 600 is taken out with interest charged at 8% p.a. Find
  a the yearly interest on this mortgage
  b the monthly repayment of the interest
  c the interest repaid over a 25-year period.

23 Interest of 9% p.a. is charged on an endowment mortgage of £8400. Find
  a the annual interest which is charged
  b the monthly interest which is repaid
  c how much interest is repaid during the 20-year life of the mortgage.

24   **a** What is the annual interest on an endowment mortgage of £11 400, if interest is charged at 12% p.a.?
  b What will the monthly repayments be?
  c How much interest is paid over the 15 years which the mortgage runs?

25   **a** Calculate how much interest is paid during one year at 13% p.a. on an endowment mortgage of £19 200.
  b How much interest is repaid each month?
  c How much interest is repaid during the 25-year span of the mortgage?

26 You want to take out an endowment mortgage for £16 800 with interest charged at 11% p.a. How much should you expect to pay
  a in one year   **b** each month   **c** over 15 years?

Mortgages

27 You want to buy a house for £26 000 and you can afford a deposit of £6600. The Building Society will charge you 9% p.a. on an endowment mortgage for the rest of the money.
  a How much do you have to borrow and how much interest will you pay on it in one year?
  b How much will the interest be (i) each month   (ii) over 20 years?

28 Mr and Mrs Benton wish to buy a house for £32 500 and they can afford a deposit of £11 200. The Building Society will loan them the remainder as an endowment mortgage at 13% p.a.
  a What is the size of their mortgage and what is the annual interest they pay?
  b What is the repayment   (i) each month   (ii) over 24 years?

29 Miss Haines wants to buy a 2-bedroomed flat for £18 650 and she asks a Building Society for an endowment mortgage of £12 840 at 8% p.a.
  a What deposit is she able to pay?
  b How much interest does she pay back each month?
  c If she also pays £34·70 each month on life insurance, what is her total monthly payment?

30 John Staines has been given an endowment mortgage of £13 560 at 11% p.a.
  a How much interest does he repay each month?
  b He also pays £27·50 on life insurance each month. What are his total monthly payments?

31 Each year David Baxendale earns £6200 and his wife Anne earns £3600. The Building Society will give them a mortgage equal to *either* two and a half times David's annual salary alone, *or* one and a half times their joint annual salaries. They have enough money to put down £4500 as a deposit on the house of their choice.
  a What is the maximum mortgage which the Building Society will give them?
  b What is the price of the most expensive house they can buy?
  c If they buy the highest-priced house they can afford, and the Building Society charges interest at 9% p.a., how much interest do they pay in the first year?
  d How much are their monthly repayments of this interest?

32 A Building Society will give Paul Turner a mortgage at 11% p.a. provided that he has a deposit of £5340 and that he asks for a mortgage no greater than twice his annual earnings.
  a He sees a house for £24 600 and he can afford the deposit. How much does he have to earn each year so that he can get a large enough mortgage?
  b If he gets the mortgage, how much interest does he repay
    (i) in his first year   (ii) each month?

33 A young couple are buying their first house. They can get a mortgage for 90% of the price of the house, provided this amount is less than twice their joint annual incomes. The other 10% they have to provide as a deposit.
  a If he earns £3700 and she earns £3400 each year, what is the maximum mortgage they can get?
  b They see a house they like for £16 500. Find 90% of this price and say if the house is too expensive or not for them.
  c They eventually buy a house for £12 600 taking out the 90% mortgage at an interest rate of 8% p.a. Find the size of their mortgage and the size of the deposit they need.
  d Find the interest they repay (i) yearly   (ii) monthly.

Mortgages

34 The Bubwith & Barlby Building Society are offering mortgages of 85% of the price of a house. The customer must provide the other 15% as a deposit and satisfy the condition that the size of the mortgage must be less than $2\frac{1}{2}$ times the husband's annual income.

  a Mr Townsend earns £7450 each year and he wants to buy a house costing £21 600. Calculate 85% of the price of the house to find whether the Building Society will give him a mortgage.

  b If he buys the house with an 85% mortgage, how much deposit is needed?

  c If interest is calculated at 8% p.a., how much interest will he pay each month?

Inflation

Part 1

1 A house costs £20 000 now, but its price increases by 10% each year.

Copy and complete the table alongside and find how much the house will cost after 4 years.

Cost now	£20 000
Increase	+__________
Cost after 1 year	
Increase	+__________
Cost after 2 years	
Increase	+__________
Cost after 3 years	
Increase	+__________
Cost after 4 years	__________

2 A car costs £4500 now, but the price of this model increases by 10% each year. Use the same type of table as above to find out how much the car will cost after 4 years.

3 A new carpet for the dining-room will cost £600 today. If its purchase price increases by 10% each year, use a table like the one above to find how much the carpet would cost to buy in 4 years time.

4 Today's price for some hi-fi equipment is £420, but its price will increase by 10% each year. Find how much it costs to buy similar equipment in 4 years time. (Give the answer to the nearest penny.)

5 A new house costs £20 000 this year, but its price increases by 8% each year.

Copy and complete this table to find its cost after 4 years. (Round off each entry in the table to the nearest whole £.)

Cost now	£20 000
Increase	+__________
Cost after 1 year	
Increase	+__________
Cost after 2 years	
Increase	+__________
Cost after 3 years	
Increase	+__________
Cost after 4 years	__________

6 A new car costs £5000 now, but inflation at 8% per annum increases its price. Find how much it would cost to buy after 4 years. (Round off the calculations to the nearest £ where necessary.)

7 A suite of furniture costs £650 today. If inflation is 8% p.a., find how much it would cost to buy in 3 years time. (Where necessary, round off the calculations to the nearest whole £.)

8 The return flight to Bombay costs £540 at today's prices. If this price inflates by 6% p.a., find how much this flight will cost in 3 years time. (Round off any calculations to the nearest whole £.)

9 Mr and Mrs Small want to extend their house by building an extra bedroom which will cost them £2800 if they do it immediately. However they decide to save up for 3 years during which time prices increase by 5% p.a. What will the extension cost them then? (Round off calculations to the nearest whole £).

Inflation

10 Mrs Tydeman has a part-time job for which she earns £1600 in a year. Her wage rises in line with inflation which runs at 12% p.a. for 4 years. Rounding off calculations to the nearest whole £, calculate her annual earnings in 4 years time.

In questions **11–19**, calculations should be rounded off to the nearest penny where necessary.

11 Find the cost in 3 years time of a bicycle with a price today of £140, if prices inflate at 6% p.a.

12 How much will a television set cost in 4 years time, if its price now is £170 and prices inflate by 8% p.a.?

13 What will be the price of a journey by train in 3 years time, if it costs £25 now and prices inflate by 12% p.a.?

14 A clarinet costs £72 now but this price increases by 15% p.a. over the next 2 years. What will it cost in 2 years time?

15 A gas fire's present price is £68 but this price inflates by 14% p.a. over a 3-year period. Find its price at the end of these 3 years.

16 What will a stair-carpet cost in 4 years time, if today's price is £185 and prices increase by 16% p.a.?

17 A builder charges Mr Smith £245 for building a wall. If Mr Smith had waited two years during which time costs increased by 15% p.a., how much would the builder have charged then?

18 The Christmas turkey cost £18 last year, but prices have inflated by 21%.
   a How much will a similar turkey cost this year?
   b If inflation continues at the same rate next year, how much will I pay for a similar turkey next Christmas?

19 Jim Blumen spent £8·50 on flowers for his wife's birthday last year.
   a If prices inflated by 12% during last year, how much will he have to spend this year for a similar bouquet?
   b If prices this year are increasing by 15%, what will he have to spend on similar flowers next year?

20 To compare the effects of different rates of inflation, imagine an article costs £1000. Find out how many years it would take before it doubles to £2000 for these rates of inflation. (Round off where necessary to the nearest whole £.)
   a 30% inflation p.a.   **b** 20% inflation p.a.
   c 15% inflation p.a.   **d** 10% inflation p.a.
   e 5% inflation p.a.

Compare the answers with each other. With inflation of 2% p.a. a similar calculation will show that it takes 34 years to double the price of the article.

Part 2

1 Last year a bicycle cost £120 but now its price is £126. By what percentage has the price increased?

2 A firm makes foodmixers costing £60, but the price is increased to £72. Find the percentage rise in the price.

Inflation

3   Last September it cost £84 to fly to Paris, now it costs £105. Find the percentage increase in the fare.

4   The return fare to Bolton by train was £25 a year ago, now it is £27. What percentage increase is this?

5   A new car cost £5760 last year. If its present price is £6048, find the percentage increase.

6   A radio cost £160 a few months ago but its price now is £172. What percentage increase is this?

7   A package-holiday was advertised at £240 in last year's brochure. The same holiday will now cost £246. Find the percentage increase in the price.

8   The Retail Price Index (RPI) stood at 126 a year ago, now it is 147. What was the rate of inflation last year?

9   If the RPI was 135 at the start of the year and 144 at the end of the year, calculate the annual rate of inflation.

10   John Sedgeworth earned £6480 last year and £6885 this year. What percentage increase in his earnings was this?

11   A colour-film cost £4·50 until the manufacturer increased the price to £4·86. What percentage increase is this?

12   Mrs Andrews has been given a rise so that her daily earnings have increased from £17·60 to £19·80. Calculate the percentage increase in her earnings.

13   A typical shopping basket of food cost £27·50 last year; the same food now costs £30·80. Find the rate of inflation of the price of food.

14   The quarterly rental charge for a telephone was £10·50 but now it has been raised to £12·00. By what percentage has the charge been increased?

15   One unit of electricity cost 3.92 pence at the beginning of the year and 4.41 pence at the end of the year. Calculate the percentage rate at which the price of electricity has risen.

Bank accounts

Part 1 Deposit accounts

Money deposited in this type of account will yield interest at a certain percentage or rate.

However, no cheque book is issued and no payment of bills can be made direct from these accounts. They can be thought of as a type of savings account.

1 Calculate the interest earned in one year on these amounts at the given rates.

a	£345 at 8% p.a.	b	£825 at 7% p.a.	c	£542 at 5% p.a.
d	£726 at 9% p.a.	e	£2425 at 6% p.a.	f	£3046 at 8% p.a.
g	£484 at 12% p.a.	h	£752 at 14% p.a.	i	£906 at 15% p.a.
j	£774 at 13% p.a.	k	£1622 at 14% p.a.	l	£2721 at 16% p.a.
m	£4915 at 17% p.a.	n	£3764 at 18% p.a.	o	£6078 at 15% p.a.
p	£12 794 at 19% p.a.				

2 Calculate (i) the interest earned in one year on these amounts
   (ii) the total sum of money in the account at the end of the year.

a	£450 at 6% p.a.	b	£840 at 5% p.a.	c	£620 at 5% p.a.
d	£950 at 8% p.a.	e	£350 at 7% p.a.	f	£570 at 6% p.a.
g	£1270 at 8% p.a.	h	£2140 at 9% p.a.	i	£725 at 12% p.a.
j	£615 at 14% p.a.	k	£326 at 15% p.a.	l	£407 at 13% p.a.
m	£1625 at 11% p.a.	n	£2422 at 16% p.a.	o	£8134 at 14% p.a.
p	£7063 at 14% p.a.				

3 Calculate (i) the interest earned in the first year on these amounts of money
   (ii) the total sum in the account at the end of the first year
   (iii) the interest earned on this total in the second year
   (iv) the new total sum in the account at the end of the second year.
This type of interest is called **compound interest**.

a	£650 at 10% p.a.	b	£840 at 10% p.a.	c	£750 at 10% p.a.
d	£460 at 5% p.a.	e	£528 at 5% p.a.	f	£844 at 5% p.a.
g	£324 at 15% p.a.	h	£908 at 15% p.a.	i	£1236 at 15% p.a.
j	£1350 at 16% p.a.	k	£2150 at 12% p.a.	l	£4850 at 14% p.a.

4 The above method of calculating the amount of money in a deposit account after several years is very long. A formula can be used instead.

If the amount invested or **principal** is £P, then the amount in the account £A after n years at a rate of R% per annum is given by the formula

$$A = P\left(1 + \frac{R}{100}\right)^n.$$

Calculate A when

a	$P = 352,\ R = 16\%,\ n = 5$		b	$P = 416,\ R = 13\%,\ n = 4$
c	$P = 223,\ R = 15\%,\ n = 7$		d	$P = 682,\ R = 12\%,\ n = 3$
e	$P = 179,\ R = 18\%,\ n = 6$		f	$P = 642,\ R = 11\%,\ n = 8$
g	$P = 704,\ R = 19\%,\ n = 5$		h	$P = 917,\ R = 9\%,\ n = 4$
i	$P = 429,\ R = 8\%,\ n = 8$		j	$P = 633,\ R = 7\%,\ n = 10$
k	$P = 1250,\ R = 16\%,\ n = 7$		l	$P = 2470,\ R = 18\%,\ n = 4.$

Bank accounts

Part 2 Current accounts

Money in this type of account earns *no* interest, and so it is not to be used for savings. However, a cheque book is issued and regular payments (of bills, mortgages, insurance policies etc.) can be made from the account by standing orders or direct debits.

1 Look at this blank cheque. Write
   a the name of the person whose account this is
   b the number of this particular cheque
   c the number of the account   **d** the code number of the bank.

2 Make three copies of the blank cheque.
Using the name Pat Williams, make out a cheque to pay
   a Mr Andrew Wallace the sum of £42·34
   b Westlands Electricity Board the sum of £85·92
   c yourself the sum of £35 in cash.

3 Copy the table in this bank statement, and for each transaction described below, make an entry onto the statement and keep a balance.

National Bank

Statement of account with National Bank plc

Description of entries
BGC Bank Giro Credit
D/D Direct Debit
S/O Standing Order

All entries to 10 FEB inclusive are complete.

MR J BRIGGS,
33 HARDUP LANE,
BROKEBY.

Account number
0123456

Date	Particulars	Payments	Receipts	Balance *When overdrawn marked OD*
19__	Opening Balance			120 : 34
10 JAN	CREDIT BY POST			
13 JAN		511245		
14 JAN		511246		
21 JAN	SUNDRY CREDIT			
26 JAN	EMPLOYERS BGC			
27 JAN		511247		
28 JAN		511248		
1 FEB	HOMELY BUILDING SOC S/O			
1 FEB	DISTRICT COUNCIL S/O			
4 FEB		511249		
7 FEB		511250		
8 FEB	INSURANCE D/D			
9 FEB		511251		
9 FEB	SUNDRY CREDIT			
10 FEB		511252		

Bank accounts

The account opens in January with a balance of £120·34.

On 10 January the bank receives cheques totalling £74·45 to put into your account.

On 13 January the bank pays cheque number 511245 for £12·52.

On 14 January the bank pays cheque number 511246 for £26·00.

On 21 January the bank receives cheques totalling £51·00 to put into the account.

On 26 January your monthly salary of £230·54 is paid in by Bank Giro Credit (BGC).

On 27 January the bank pays cheque number 511247 for £35·00.

On 28 January the bank pays cheque number 511248 for £11·50.

On 1 February your mortgage of £82·10 is paid by Standing Order (S/O).

On 1 February your rates of £15·04 are paid by Standing Order (S/O).

On 4 February the bank pays cheque number 511249 for £42·31.

On 7 February the bank pays cheque number 511250 for £18·12.

On 8 February your insurance for £11·42 is paid by Direct Debit (D/D).

On 9 February the bank pays cheque number 511251 for £92·10.

On 9 February the bank receives cheques totalling £11·52 to put into your account.

On 10 February the bank pays cheque number 511252 for £10·23.

4 Copy the table on this next bank statement, and for each transaction given, make an entry.

Keep a balance but note when the account is *overdrawn*, the letters OD are used in the balance column, rather than a negative sign.

National Bank	Statement of account with National Bank plc	Description of entries BGC Bank Giro Credit D/D Direct Debit S/O Standing Order
All entries to 2 JUNE inclusive are complete.	MR J BRIGGS, 33 HARDUP LANE, BROKEBY.	Account number 0123456

Date	Particulars		Payments	Receipts	Balance *When overdrawn marked OD*
19—	Opening Balance				15 : 84
5 MAY		701831			
6 MAY		701832			
6 MAY	CREDIT BY POST				
9 MAY		701833			
9 MAY		701834			
10 MAY	SUNDRY CREDIT				
16 MAY	HOMELY BUILDING SOC S/O				
26 MAY	EMPLOYERS BGC				
27 MAY	INSURANCE D/D				
27 MAY	BROKEBY DISTRICT COUNCIL S/O				
30 MAY		701835			
30 MAY		701836			
2 JUNE	CREDIT BY POST				

The account opens in May with a balance of £15·84.

On 5 May the bank pays cheque number 701831 for £8·50.

On 6 May the bank pays cheque number 701832 for £5·10.

On 6 May the bank receives cheques for £14·76 to put into your account.

On 9 May the bank pays cheque number 701833 for £18·00.

Bank accounts

On 9 May the bank pays cheque number 701834 for £4·00.
On 10 May the bank receives cheques for £65·00 to put into your account.
On 16 May your mortgage of £62·00 is paid by Standing Order (S/O).
On 26 May your salary of £148·10 is paid into your account by Bank Giro Credit.
On 27 May your insurance for £32·65 is paid by Direct Debit (D/D).
On 27 May your rates of £14·20 are paid by Standing Order (S/O).
On 30 May the bank pays cheque number 701835 for £87·25.
On 30 May the bank pays cheque number 701836 for £20·00.
On 2 June the bank receives a cheque for £14·62 to put into your account.

5 Construct your own table for a bank statement, so that it shows the payments, receipts and balance for these transactions.

The opening balance should be £247·52. Remember to indicate an overdrawn account by using the letters OD in the balance column.
On 26 September the bank receives your monthly salary of £565·27 by Bank Giro Credit.
On 29 September your rates of £31·40 are paid by standing order to the local council.
On 3 October your mortgage of £124·65 is paid by standing order.
On 4 October your insurance premiums for £31·66 are paid by direct debit.
On 5 October the bank receives £64·10 to credit to your account.
On 7 October the bank pays cheque number 243361 for £347·82.
On 10 October the bank pays cheque number 243359 for £194·10.
On 11 October you withdraw £40 in cash using cheque number 243362.
On 12 October the bank pays cheque number 243360 for £117·26.
On 13 October the bank pays cheque number 243363 for £12·60.
On 13 October the bank receives £24·70 to credit to your account.
On 19 October you withdraw £15 in cash using cheque number 243364.
On 27 October the bank receives your monthly salary of £565·27 by Bank Giro Credit.

Hire-purchase

Part 1

1 Some hi-fi equipment can be bought either for £246 cash or on hire-purchase with a deposit of £25 followed by eight monthly payments of £31·50. Find
   a   the total hire-purchase price
   b   the extra paid over the cash price.

2 A bicycle can be bought on hire-purchase with a deposit of £19 followed by six monthly instalments of £21·50. If the cash price is £132, find
   a   the total hire-purchase price
   b   the extra paid over the cash price.

3 A young couple see a gas cooker advertised at £214 cash. They decide however to buy it on H.P. with a deposit of £35 and twelve weekly instalments of £17.
   a   What is the total hire-purchase price?
   b   How much extra will they have paid by buying on H.P. rather than with cash?

4 A pair of binoculars is advertised at a cash price of £65. A shop has them on offer for £9 deposit and five weekly payments of £12·34.
   a   What is the total H.P. price?
   b   How much extra is paid by buying on H.P. rather than with cash?

5 A camera is advertised at £174 cash. It can also be bought on hire-purchase for a deposit of £25 followed by fifteen weekly instalments of £10·45.
   a   What is the hire-purchase price of the camera?
   b   How much is saved by paying cash for it?

6 An electric typewriter can be bought either for £206 cash or on H.P. with a deposit of £24 followed by sixteen weekly payments of £13·95.
   How much cheaper is it to buy the typewriter with cash rather than on H.P.?

7 A three-piece suite can be bought on hire-purchase in two different ways.
   The *short-term* method requires a deposit of £40 followed by twelve monthly payments of £27·83.
   The *long-term* method needs a £35 deposit followed by twenty-four monthly payments of £14·95.
   Find the cost of the furniture under both these schemes. Why is one scheme more expensive than the other?

8 A holiday in Spain is advertised at a cash price of £145 per person. It can also be paid for on H.P. with a deposit of one fifth of the cash price followed by twelve weekly instalments of £10·55.
   a   Find the cost of the holiday when paid for on hire-purchase.
   b   How much cheaper is it to pay cash?

9 Gary Dodworth bought a radio/cassette recorder on H.P. with a deposit of one quarter of the cash price followed by sixteen weekly instalments of £3·42. If the cash price is £65·40, find
   a   the total hire-purchase price
   b   how much more expensive it was on H.P.

Hire-purchase

10 A colour television was advertised at £284 cash. Angela Hemingway bought it on H.P. with a deposit of one eighth of the cash price and nine monthly payments of £29·62. How much more expensive is it on H.P.?

11 Mr and Mrs Watkinson want to buy a bedroom carpet 7 metres long and 6 metres wide. They decide on one with a cash price of £6·45 per square metre. Hire-purchase terms offered require a deposit of one fifth of the cash price followed by one year's monthly payments of £20·35. Find
   a   the cash price of the carpet they wish to buy
   b   the H.P. price of the carpet
   c   the increase in the cost, if they buy it on H.P.

12 John Oldroyd bought his wife a fridge-freezer for Christmas on hire-purchase. The cash price was £189·60 and he paid a deposit of one sixth of the cash price followed by fourteen weekly instalments of £12·56. How much more expensive was it to buy it on these H.P. terms?

13 A firm sells £254-worth of kitchen units on hire-purchase. The deposit is 12% of the cash price and is followed by fifteen weekly instalments of £17·46. Calculate
   a   the deposit   b   the total H.P. price
   c   how much cheaper it is to pay cash.

14 Mr and Mrs Shaw are quoted a cash price of £946 to have their house centrally heated. They decide to pay for it on hire-purchase, the terms of which are a deposit of 15% of the quoted price, followed by monthly instalments of £39·80 over two years. Calculate
   a   the deposit   b   the total H.P. price
   c   how much more expensive it is for them on H.P.

15 To double-glaze their entire house, Mr and Mrs Booth are quoted a price of £1485. To pay on hire-purchase, the terms would be a deposit of 18% of this quoted price followed by monthly payments of £43·65 over a period of three years. Calculate
   a   the deposit   b   the total H.P. price
   c   how much cheaper it is for them to pay cash.

16 A second-hand car is advertised at £3250 cash. Marianne Allsopp decides to buy it but she has only £950 in cash. She can borrow the difference required from a bank which will charge her 16% p.a. interest, if she repays the amount in one year. Otherwise, she can buy the car on hire-purchase where the deposit will be 22% of the cash price followed by monthly payments of £86·45 over three years. Find
   a   the interest the bank would charge, and the total cost of the car, if she borrowed from the bank
   b   the deposit required for H.P. and the total cost of the car, if she bought it on H.P.
   Which method of purchasing the car would you recommend?

Hire-purchase

Part 2

1 The total cost of a table and chairs on hire-purchase is £116·90. This is paid partly by an initial deposit and partly by twelve weekly instalments of £7·45 each. How much is paid as a deposit?

2 A young couple buy a new dining-room carpet on hire-purchase for £386·10 by paying a deposit followed by fifteen weekly payments of £22·74.
What amount did they pay as a deposit?

3 A single-lens reflex camera costs £194·34 on H.P., and can be bought by placing a deposit which is followed by sixteen weekly instalments of £10·24.
How much is the deposit?

4 A travel agent is offering H.P. terms for a two-week holiday in Greece costing £345. He requires a deposit and eighteen weekly payments of £16·25.
What deposit does the travel agent require?

5 Mrs Mayo buys herself a wool coat on H.P. at a total cost of £64·12. If she pays a deposit of £9·40 and there are eight equal weekly payments to follow, how much will each of them be?

6 Mr and Mrs Taynton pay a total of £472 on hire-purchase for the installation of a shower in their bathroom. They pay a deposit of £79 followed by six equal monthly payments. What amount do they pay each month?

7 The total price of a second-hand car on H.P. is £1270, where a deposit of £205 is needed together with twelve equal monthly instalments.
Find how much must be paid each month.

8 A sales representative arranges for Mr Farrow to pay for two solar panels in his house roof on hire-purchase. The total cost will be £1420, part of which is paid as a deposit of £187, and the rest is paid in fifteen equal monthly instalments. How much is each instalment?

9 Judy Simms' engagement ring cost £193. It was bought on H.P. with a deposit of £25 and weekly payments of £12 each.
How many weeks will it take to pay for the ring?

10 Mrs Fosdyke bought an electric cooker on hire-purchase for £369. She had to pay an initial deposit of £45 followed by several weekly instalments of £21·60 each. For how many weeks did she have to pay these instalments?

11 A builder's van can be bought second-hand for £2780 on hire-purchase with an initial downpayment of £350 followed by monthly instalments of £97·20 each. How long will it take to pay for the van?

12 An antique bookcase is sold for £846 on hire-purchase. The deposit is £165·60 and the monthly payments are £32·40 each.
How long does it take to pay off this debt?

13 An extension to a house costs £4263 and is paid for on H.P. with an initial deposit followed by monthly instalments of £101·50 over 28 months.

   a Find the initial deposit.

   b Express this deposit as a fraction of the total cost of the extension.

Hire-purchase

14 A slide projector is offered on hire-purchase for a deposit followed by twelve weekly payments of £4·48 each. If the total cost of the projector is £67·20, find
   a   the deposit required
   b   the deposit as a fraction of the total cost.

15 Mrs Mortimer buys a wedding present costing £75 on hire-purchase for her niece. If the deposit is followed by fifteen weekly payments of £3·25 each,
   a   find the deposit
   b   express the deposit as a percentage of the total cost of the present.

16 A food mixer is advertised at a price of £94·50 on H.P. An initial down payment and fourteen weekly instalments of £5·13 each are needed.
   a   Find the initial down payment.
   b   Express it as a percentage of the total cost.

Reading the meter

Write the readings given by these sets of meters.

1 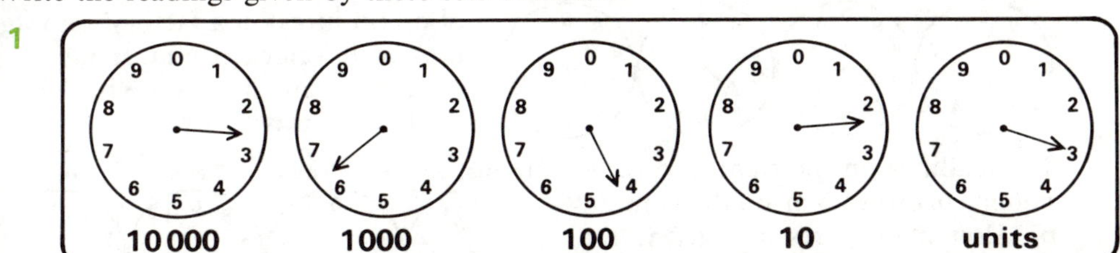

2 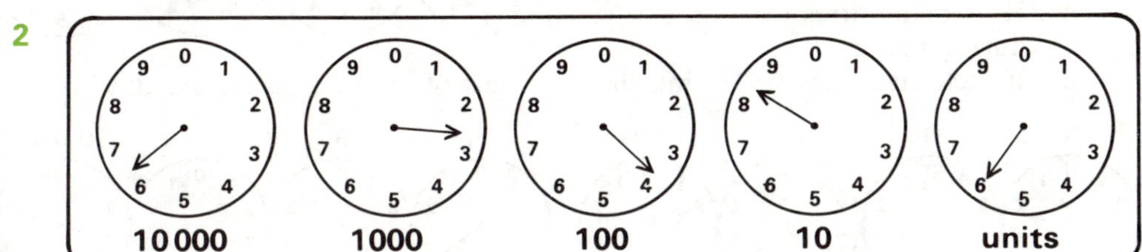

3 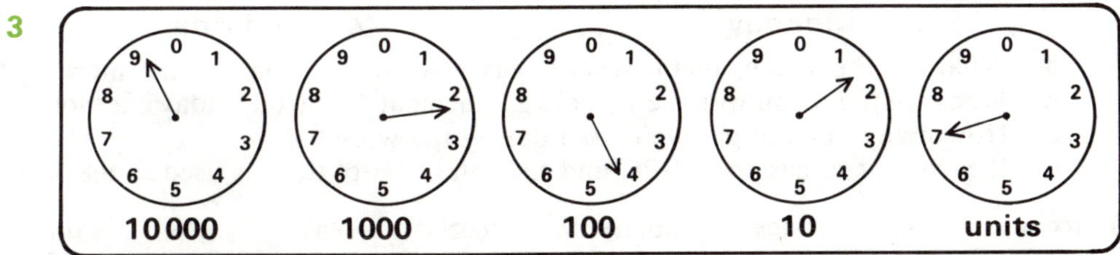

4 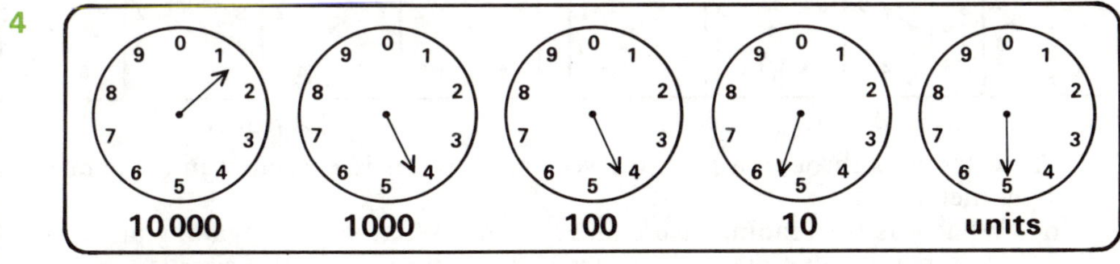

5 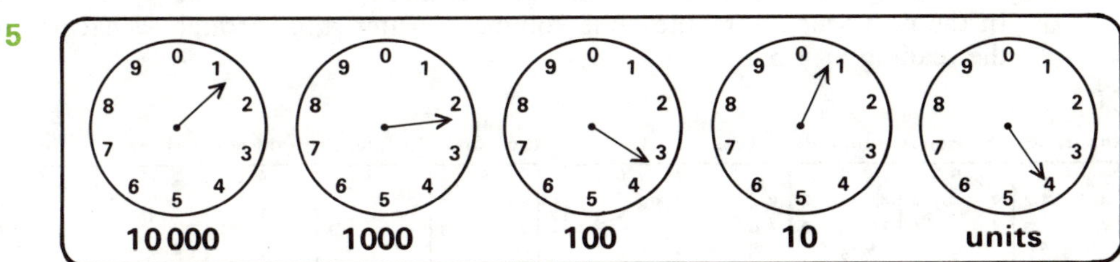

6 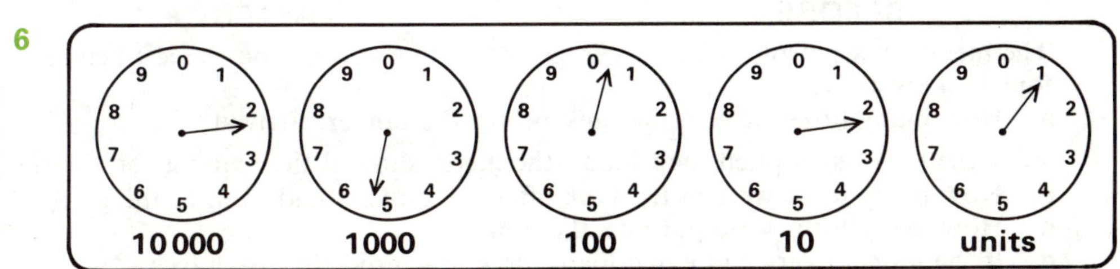

Reading the meter

7

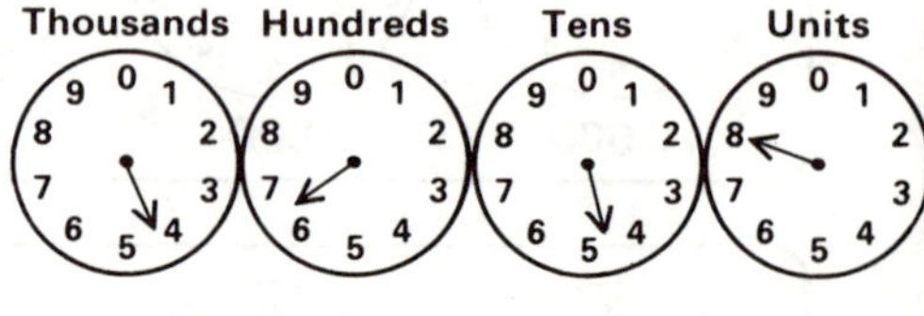

The readings here give the volume of oil (in litres) in a factory's storage tank at 8 o'clock one morning.

a How many litres were in the tank at this time?

The readings on the right give the volume of oil at 5 p.m. on the same day.

b How many litres were left in the tank at 5 p.m.?

c How many litres were used during the day?

d If each litre costs £0·36, find the total cost of oil used during the day.

8

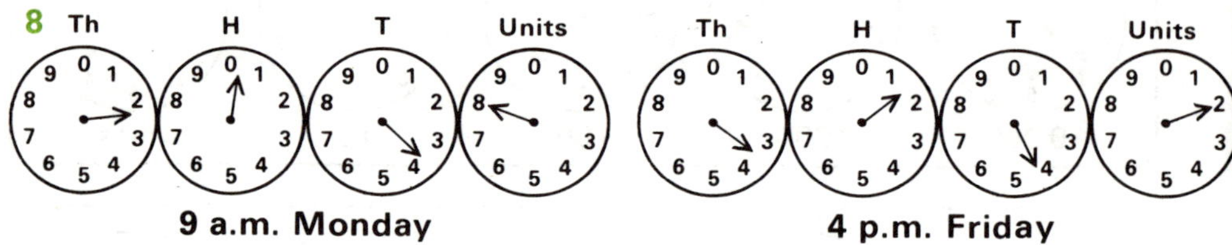

9 a.m. Monday   **4 p.m. Friday**

a What was the reading on the school's gas meter at 9 a.m. on Monday morning?

b What was the reading on the school's gas meter at 4 p.m. on Friday afternoon?

c How many units of gas were used during the week?

d If each unit of gas cost £0·26, find the total cost of the gas used in the week.

9

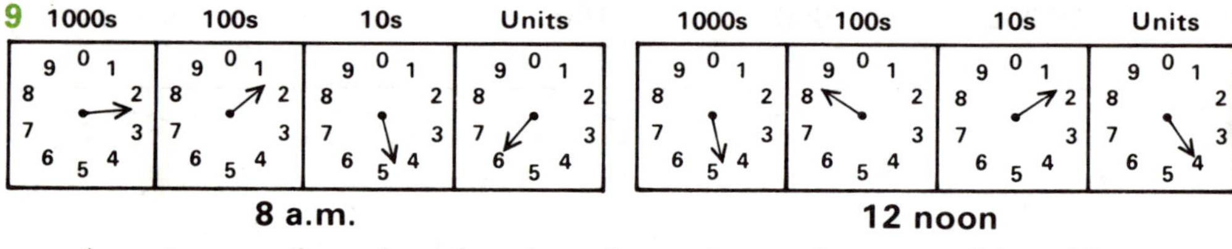

8 a.m.   **12 noon**

As water runs through a pipe, the volume of water is measured in cubic metres by a meter.

a What was the reading at 8 a.m.?   **b** What was the reading at 12 noon?

c How many cubic metres of water had run through in this time?

d In the next four hours, the same volume of water runs through. What will be the reading at 4 p.m.?

10

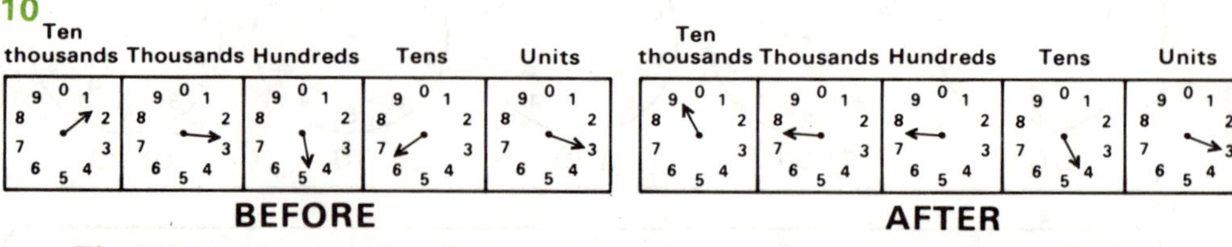

BEFORE   **AFTER**

The meter on a petrol tank at a garage shows the readings on the **left** before four tankers arrive.

a How many litres were in the tank before the tankers arrived?

After the tankers emptied their loads, the meter showed the readings on the **right**.

b How many litres were in the tank after the tankers had been to the garage?

c How many litres were put into the tank?

d If the four tankers had equal loads, how much did they each carry?

Reading the meter

11   **A**   **B**

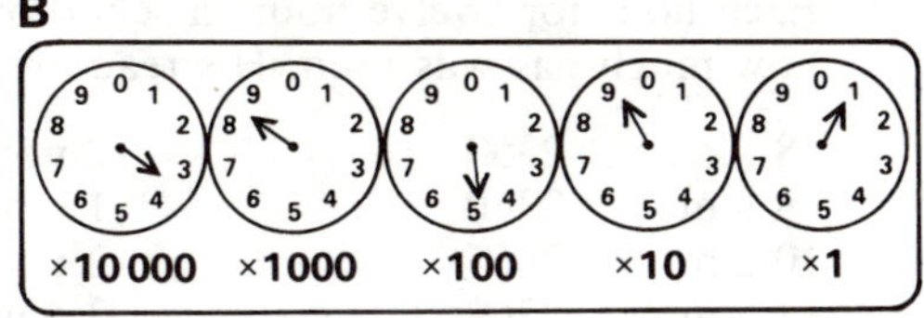

At 9 a.m., the gas meter in a factory building gave reading **A**. By 3 p.m. of the same day the meter gave reading **B**.

a   How many units of gas had been used?
b   How many hours had it taken to use this gas?
c   What was the average rate of gas used in units per hour?

12   **Y**   **Z**

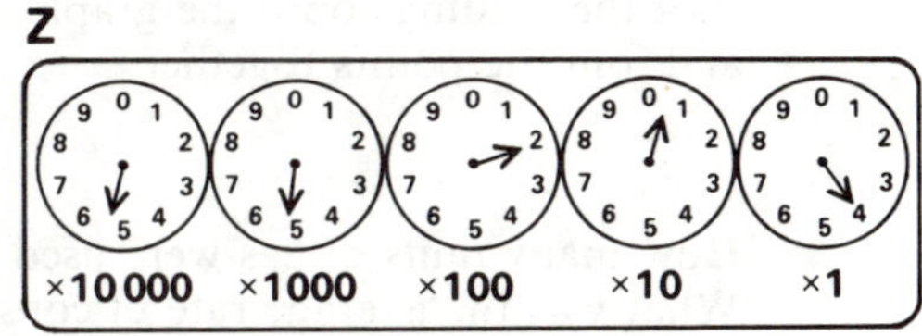

An office-block has oil-fired central heating. At 8 a.m. the fuel tank gave the reading **Y** in litres. At 5 p.m. the same day the reading **Z** was taken.

a   How many hours had the central heating been on for?
b   How many litres of oil had been used?
c   What was the average rate of oil used in litres per hour?

13   **P**   **Q**

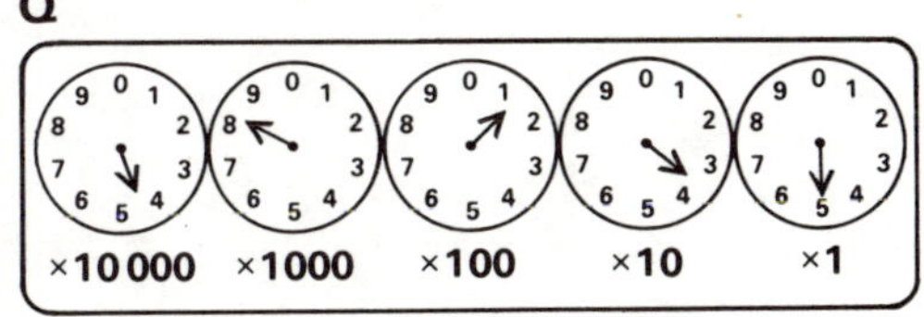

The meter on the water tank in the roof of a multi-storey building gave reading **P** at 10 a.m. and one hour later it gave reading **Q**.

a   How many litres of water were used in this time?
b   If the water was used at a steady rate, how much was used per minute?
c   How much was used per second?
d   If in the next hour only half as much water was used, what would the reading on the metre be at 12 noon?

14   **6 a.m.**   **2 p.m.**

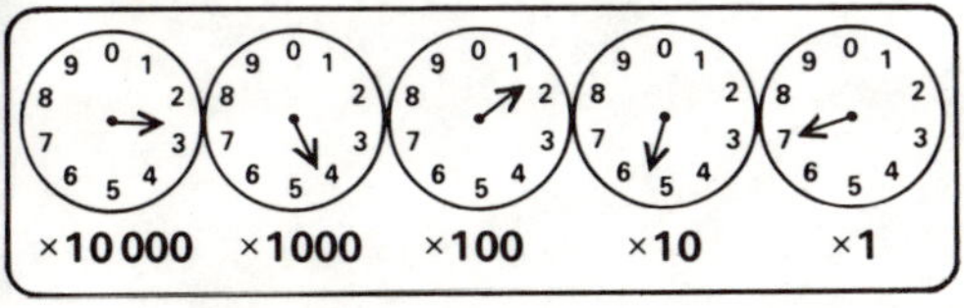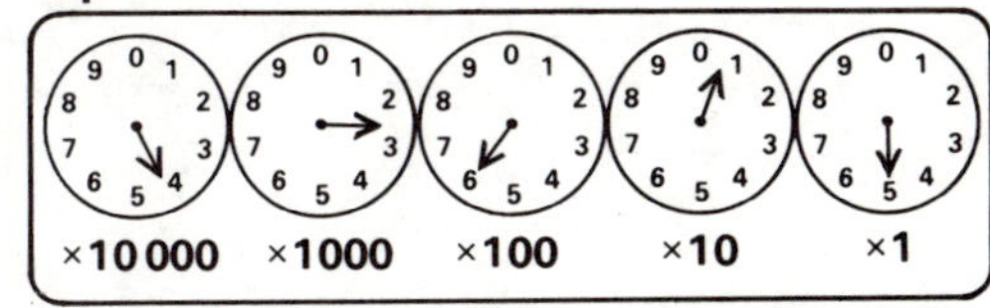

Coal mixed with water is transported in a slurry along a pipeline. The meter, measuring the flow in cubic metres, gives the readings at the start and end of the morning shift as shown above.

a   How many cubic metres of slurry were carried along the pipeline during this shift?
b   What was the average amount of slurry transported per hour?
c   If each cubic metre of slurry contains 250 kg of coal, what mass of coal was transported per hour?
d   If 1000 kg = 1 tonne, give the previous answer in tonnes.

Reading the meter

15 Each hour for twelve hours a schoolboy read the gas meter in his home to find how much gas was used. His readings are given here.

8 a.m.	2350	12 noon	2373	4 p.m.	2398
9 a.m.	2360	1 p.m.	2390	5 p.m.	2402
10 a.m.	2365	2 p.m.	2392	6 p.m.	2428
11 a.m.	2369	3 p.m.	2395	7 p.m.	2440
				8 p.m.	2450

a Draw axes, labelling the time axis from 8 a.m. to 8 p.m. and the other axis from 2350 to 2450.

Plot the readings onto the graph and join the points together.

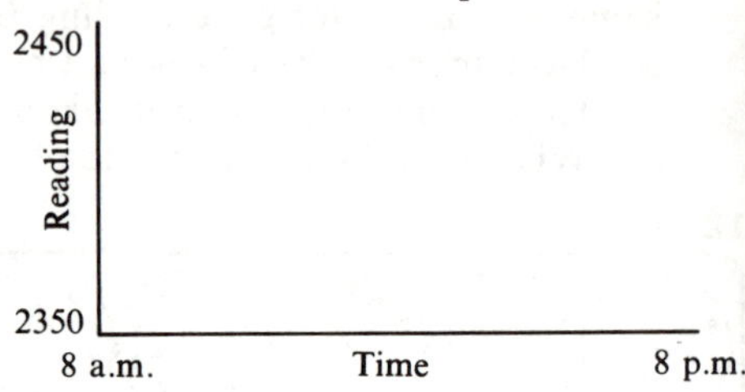

b How many units of gas were used during the 12-hour period?
c What was the average rate of consumption of gas in units per hour?
d Between which two times was most gas used?
e Between which two times was least gas used?
f Can you suggest what might have been using the gas at those times when most gas was being consumed?

Household bills

Part 1 Gas bills

1 Find the number of units of gas used from the present and previous readings below.

	Present reading	Previous reading			Present reading	Previous reading
a	3849	2317		b	6205	5013
c	8386	6144		d	3705	3192
e	4163	3872		f	9237	8658
g	2035	1164		h	4207	3028
i	5014	4825		j	6138	4299

2 The first 52 therms of gas which are used cost 36 pence each. Every therm after this costs 32 pence.
(A therm is a measured amount of heat dependent on the quality of the gas.)
Find the cost of using these numbers of therms. Give each of the answers in £.
   a 60 therms   b 72 therms   c 152 therms   d 89 therms
   e 124 therms   f 179 therms   g 207 therms   h 368 therms
   i 50 therms   j 38 therms

3 The price of a therm is not always a whole number of pence. Work out the cost of these numbers of therms, rounded down to the nearest penny.
Each of the first 52 therms costs 36.4 pence, and each therm after this costs 32.2 pence.
   a 55 therms   b 62 therms   c 64 therms   d 164 therms
   e 92 therms   f 267 therms   g 324 therms   h 156 therms
   i 48 therms   j 32 therms

4 Mr Earnshaw lives alone and does not use much gas. His bill shows that he used 45 therms, each costing 36.4 pence and there is also a standing charge of £9·90 to pay.
   a How much will he pay for the gas he has used?
   b What will his total bill come to?

5 Mr and Mrs Ridge have gas central heating and their bill shows that they have used 244 therms of gas. The first 52 therms cost 36.4 pence each, after that each therm costs 32.2 pence. A standing charge of £9·90 is also made. Find
   a how much they pay for the gas used
   b the total bill.

6 Each week Mrs Reynolds puts £3 on one side for her gas bill. After 13 weeks she receives the bill which states she has used 128 therms costing 36p each for the first 52 therms, then 32p for each therm above that.
   a How much does the gas cost?
   b What will the total bill be, if there is also a standing charge of £10·50?
   c Has she saved enough to pay the bill and if not, how much more does she need?

7 In the summer, you used 158 therms of gas; and in the winter you used 417 therms. If the price is fixed at 36p per therm for the first 52 therms and then 32p per therm thereafter, how much more do you pay for your gas in winter than in summer?

Household bills

8   Mr Illingworth's gas bill includes a standing charge of £10·20 and shows that he
has used 243 therms, costing 38p each for the first 52 therms and 29p each thereafter.
The total bill comes to £86·70, but Mr Illingworth queries this, saying that he
has been overcharged.
Find   a   the cost of the gas he has used
             b   the total bill
             c   whether he is right to query the total
and, if so,   d   how much he has been overcharged.

9   The Gas Board offers two tariffs for paying a gas bill.
TARIFF A   32p for each therm used
TARIFF B   38p for the first 52 therms and then 26p per therm thereafter
Mr and Mrs Speeton use 264 therms. Which tariff would you suggest they
choose?

10   These people are offered the choice between Tariff A and Tariff B. With as little
working as possible, find out which tariff they should each choose.
   a   Mr Little used only 48 therms.   b   Mr Marcle used 104 therms.
   c   Mr Bentley used 367 therms.   d   Mrs Smithson used 65 therms.
   e   Mr Whitworth used 106 therms.   f   Miss Rothery used 138 therms.

Part 2

Gas bills depend on the strength of the gas used (its **calorific value**); some types of
gas are hotter than others. The amount of heat the gas gives out is measured in **therms**.
The total bill is in three parts:
 (i)   a charge for the first 52 therms used
 (ii)   a charge for any extra therms used
(iii)   a standing charge.
No VAT is added to a gas bill.

1   A gas meter measures the volume of gas used in **hundreds of cubic feet**.
Normally the meter is read every 3 months (quarterly). Find how many
hundreds of cubic feet are used for these pairs of readings.

	Present reading	Previous reading			Present reading	Previous reading
a	7824	7513	b		8614	8502
c	5362	5181	d		2307	2114
e	6689	6592	f		5213	4807
g	9192	8876	h		1389	0617
i	8016	7682	j		0371	0047

2   The number of **therms** used in a house depends on
 (i)   the volume of gas used
(ii)   the strength of hotness of the gas (its **calorific value**).

The number of **therms** used is given by a formula printed on the back of the bill.

$$\text{No. of therms} = \text{hundreds of cu.ft} \times \frac{\text{calorific value}}{1000}$$

Find the number of **therms** used (to the nearest whole number), if
a   the calorific value = 1560 and 275 hundreds of cu.ft are used
b   the calorific value = 1560 and 350 hundreds of cu.ft are used
c   the calorific value = 1560 and 120 hundreds of cu.ft are used

Household bills

   d   the calorific value = 1027 and 250 hundreds of cu.ft are used

   e   the calorific value = 1027 and 512 hundreds of cu.ft are used

   f   the calorific value = 1027 and 84 hundreds of cu.ft are used.

3 The first 52 therms cost 36 pence each, any further therms cost 32 pence each.
Find the total cost in £ when a household uses

a 177 therms	**b** 265 therms	**c** 306 therms	**d** 418 therms
e 554 therms	**f** 1248 therms	**g** 65 therms	**h** 54 therms
i 50 therms	**j** 47 therms.		

4 The price charged is usually *not* a whole number of pence. The total cost is then
rounded *down* to the nearest penny.
If the first 52 therms cost 36.4p each and further therms cost 32.3p each, find
the total cost in £ when a household uses

a 164 therms	**b** 275 therms	**c** 367 therms	**d** 663 therms
e 706 therms	**f** 1265 therms	**g** 64 therms	**h** 53 therms
i 45 therms	**j** 32 therms.		

5 **a** Two successive meter readings are 6812 and 6916. Find the volume of gas
used.

    b If the calorific value of the gas is 1250, use the formula on page 162 to work
out the number of **therms** used.

    c The first 52 therms cost 36.4p each. Find the charge for these.

    d All other therms cost 32.2p each. Find the charge for these.

    e How much is the cost of all the gas used?

    f A standing charge of £9·90 is now added. What is the total bill?

6 Check this gas bill as follows

    a check item **C** by subtracting **B** from **A**

    b check item **D** by using the formula with items **C** and **K**

    c check item **F** by subtracting **E** from **D**

    d check item **G** by multiplying **E** by 36.4 and rounding down

    e check item **H** by multiplying **F** by 32.3 and rounding down

    f check the total **J** by adding **G**, **H** and **I** together.

MID WEST GAS   P.O. BOX NO. 31, GRANBY, GR2 4EP.
(A part of The British Gas Corporation)

REFERENCE FOR ENQUIRIES	TARIFF	CALORIFIC VALUE	DATE OF READING	DATE OF ACCOUNT (TAX POINT)
6200000/z/2	DOMESTIC	1250 **K**	29 SEPT	31 SEPT

MR J. TURNER,
16 COOMBE DRIVE,
BARSDALE,
WORCS.

METER READINGS		GAS SUPPLIED			CHARGES	VAT		TOTAL AMOUNT
PRESENT / PREVIOUS E = ESTIMATED S = SELF READ CARD	CUBIC FEET (HUNDREDS)	THERMS	PENCE PER THERM	£		CODE	CHARGES £	£
7124 **A** 6848 **B**	276 **C**	345 **D**						
		52 **E**	36.4	18.92 **G**			0.00	18.92 **G**
		293 **F**	32.3	94.63 **H**			0.00	94.63 **H**
STANDING CHARGE				9.90			0.00	9.90 **I**
TOTAL DUE								123.45 **J**

Household bills

7 Check this gas bill in the same way.

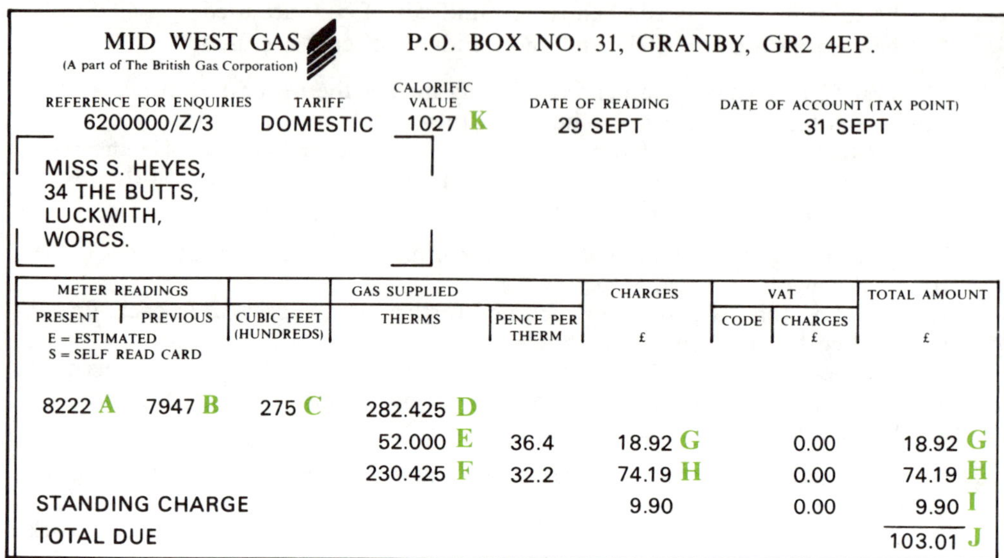

METER READINGS		GAS SUPPLIED			CHARGES	VAT		TOTAL AMOUNT
PRESENT E = ESTIMATED S = SELF READ CARD	PREVIOUS	CUBIC FEET (HUNDREDS)	THERMS	PENCE PER THERM	£	CODE	CHARGES £	£
8222 A	7947 B	275 C	282.425 D					
			52.000 E	36.4	18.92 G		0.00	18.92 G
			230.425 F	32.2	74.19 H		0.00	74.19 H
STANDING CHARGE					9.90		0.00	9.90 I
TOTAL DUE								103.01 J

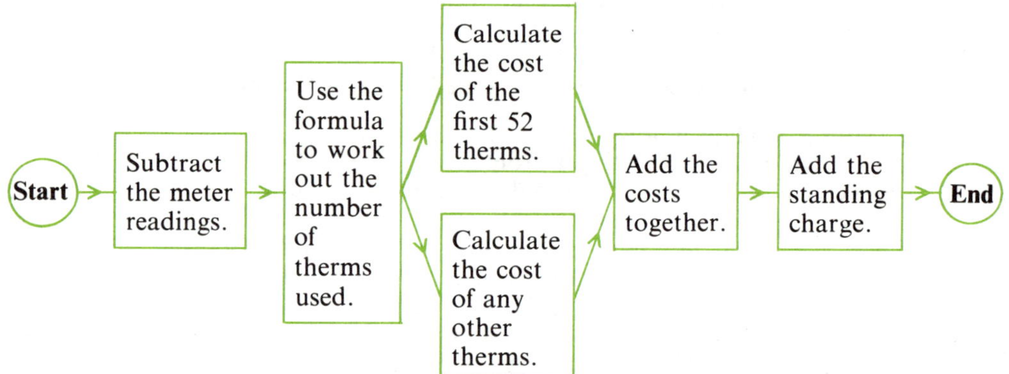

8 Use this information to calculate the total due on a gas bill. The flow diagram above will help.

Previous meter reading	3814
Present meter reading	4026
Calorific value	1250
Cost per therm for the first 52 therms	38p
Cost per therm thereafter	32p
Standing charge	£9·90

9 Calculate the total gas bill from this information.

Previous meter reading	6818
Present meter reading	7210
Calorific value	1125
Cost per therm for the first 52 therms	39p
Cost per therm thereafter	31p
Standing charge	£10·50

Household bills

10 Calculate the total gas bill from this data. Round down the cost of each item to the nearest penny where necessary.

Previous meter reading	5516
Present meter reading	5932
Calorific value	1027
Cost per therm for the first 52 therms	42.8p
Cost per therm thereafter	35.3p
Standing charge	£11·20

Part 3 Electricity bills

Electricity bills are made up of   (i) a standing charge
(ii) a charge on the Domestic Supply Tariff
and possibly (iii) a charge on the Off-peak Supply Tariff.
No VAT is added to electricity bills.

1 Each unit used on the Domestic Supply Tariff costs 5.25 pence.
Find the number of units used and their cost in £ for each of these sets of readings.

	Previous reading	Present reading			Previous reading	Present reading
a	24013	24625		b	37872	38396
c	56501	56933		d	40262	41066
e	18845	19785		f	65708	66932
g	88115	89571		h	34017	35077

2 If the charge does not work out to an exact penny, then it is rounded *down* to the nearest penny.
Take the cost per unit as 5.17 pence, and find the charge made for these sets of readings.

	Previous reading	Present reading			Previous reading	Present reading
a	42104	42347		b	56217	56672
c	10625	11468		d	33316	34218
e	61307	62632		f	80225	81286
g	12679	13666		h	26144	27323

3 Off-peak electricity is charged at a cheaper tariff of 1.9 pence per unit. It needs a special meter which has an extra standing charge of £2·50.

If 380 units are used, find
a   the cost of the units used

b   the total cost for units and standing charge.

4 a   If 563 units of Off-peak electricity are used, costing 1.9p per unit, find the charge made for them (rounding down to the nearest penny).

b   If a standing charge of £2·50 is also made, what is the total cost?

Household bills

5 Check the calculations on this bill as follows
  a check item **C** by subtracting **A** from **B**
  b check item **E** by multiplying **C** by 5.17
  c check item **F** by adding items **D** and **E**.

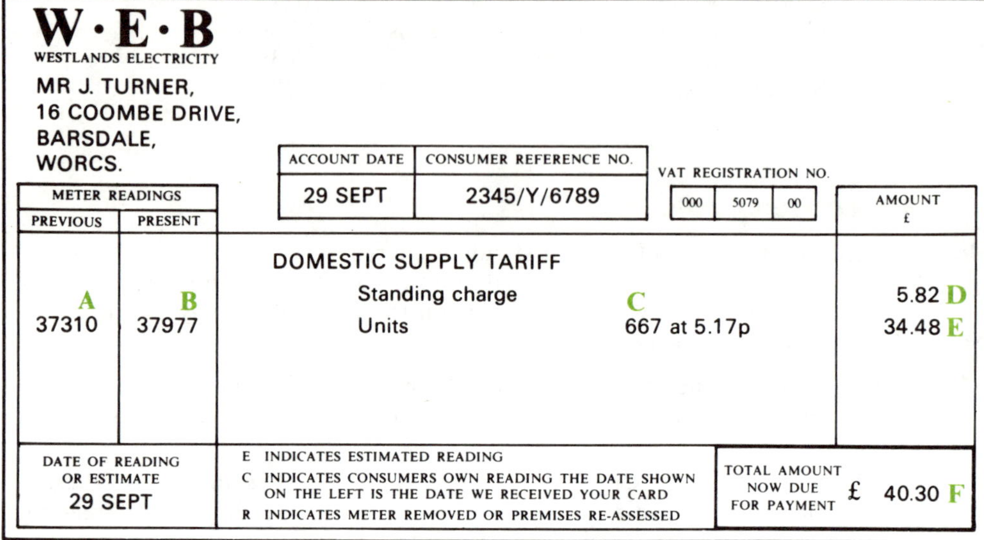

6 This bill gives readings and charges on both Domestic and Off-peak Supply Tariffs.
Check the calculations as follows
  a check item **S** by subtracting **Q** from **R**
  b check item **T** by multiplying **S** by 5.17
  c check item **X** by subtracting **V** from **W**
  d check item **Y** by multiplying **X** by 1.9
  e check item **Z** by adding **P**, **T**, **U** and **Y**.

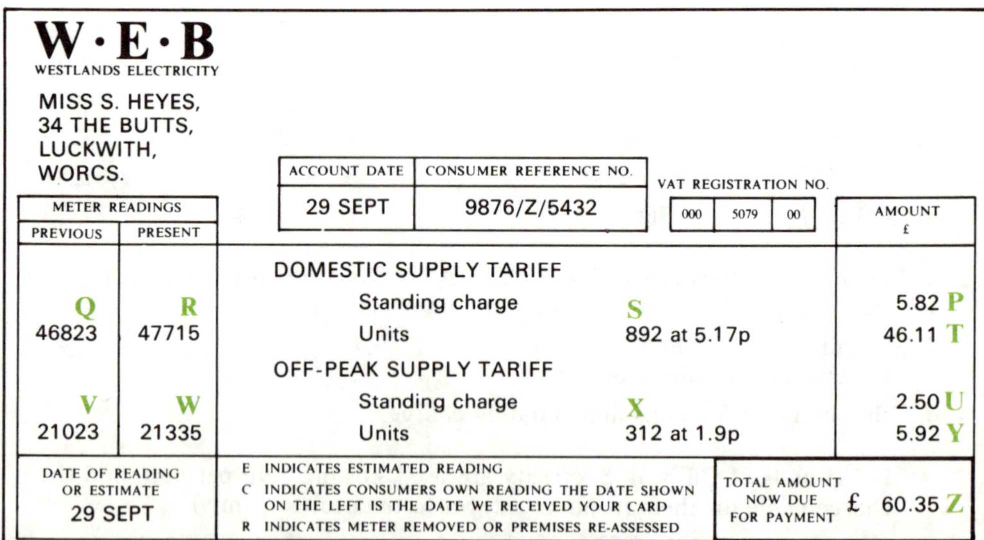

Household bills

7 Find the total amount of the electricity bill from this information.

DOMESTIC SUPPLY TARIFF   Standing charge £5·82
     857 units at 5.17p each

OFF-PEAK SUPPLY TARIFF   Standing charge £2·50
     108 units at 1.9p each

8 Find the total amount of the electricity bill from this information.

DOMESTIC SUPPLY TARIFF   Standing charge £6·10
     present meter reading   37312
     previous meter reading 36828   at 5.25p per unit

OFF-PEAK SUPPLY TARIFF   Standing charge £2·80
     present meter reading   23969
     previous meter reading 23661   at 2.25p per unit

9 Calculate what the total bill will be from this data.

DOMESTIC SUPPLY TARIFF		OFF-PEAK SUPPLY TARIFF	
Standing charge	£6·50	Standing charge	£3·10
Present meter reading	56243	Present meter reading	24597
Previous meter reading	55827	Previous meter reading	23815
Cost per unit	6.15p	Cost per unit	2.76p

10 Calculate what the total bill will be from this data.

DOMESTIC SUPPLY TARIFF		OFF-PEAK SUPPLY TARIFF	
Standing charge	£6·65	Standing charge	£3·25
Present meter reading	82198	Present meter reading	10427
Previous meter reading	81313	Previous meter reading	08913
Cost per unit	6.25p	Cost per unit	2.72p

11 Copy the basic outline of this blank electricity bill.
Fill in each missing item, starting with these meter readings, and finally work
out the total amount due for payment.

DOMESTIC SUPPLY TARIFF		OFF-PEAK SUPPLY TARIFF	
Previous meter reading	42613	Previous meter reading	22158
Present meter reading	44257	Present meter reading	22818
Standing charge	£5·50	Standing charge	£2·50

W·E·B
WESTLANDS ELECTRICITY

METER READINGS		ACCOUNT DATE	CONSUMER REFERENCE NO.	VAT REGISTRATION NO.			AMOUNT £
PREVIOUS	PRESENT	29 SEPT		000	5079	00	
		DOMESTIC SUPPLY TARIFF					
		Standing charge					...
...	...	Units ... at 5.17p					...
		OFF PEAK-SUPPLY TARIFF					
		Standing charge					...
...	...	Units ... at 1.9p					...

DATE OF READING OR ESTIMATE 29 SEPT	E INDICATES ESTIMATED READING C INDICATES CONSUMERS OWN READING THE DATE SHOWN ON THE LEFT IS THE DATE WE RECEIVED YOUR CARD R INDICATES METER REMOVED OR PREMISES RE-ASSESSED	TOTAL AMOUNT NOW DUE FOR PAYMENT £ ...

Household bills

12 Copy the basic outline of the blank bill and use this data to complete the missing items.

DOMESTIC SUPPLY TARIFF		OFF-PEAK SUPPLY TARIFF	
Previous meter reading	70143	Previous meter reading	38126
Present meter reading	71517	Present meter reading	40842
Standing charge	£5·75	Standing charge	£2·75

Remember to round down the charges to the nearest penny.

Part 4 Telephone bills

Telephone bills are made up of   (i)   a quarterly rental
      (ii)   the cost of dialled units
      (iii) calls via operator
      (iv) value added tax.

1 If each unit of time on a private line costs 3 pence, find the costs in £ of
  a 248 units   **b** 385 units   **c** 806 units
  d 1074 units   **e** 1567 units   **f** 2105 units.

2 If the quarterly rental charge is £8·25 and 567 units (at 3 pence each) are used in this quarter, find
  a the cost of the units used   **b** the total cost of units and rental.

3 687 units (at 3p each) are used in a quarter of the year when the rental is £8·25. Find
  a the cost of the units   **b** the total cost of units and rental.

4 375 units (at 3p each) are dialled in a quarter from January to April. The rental is £8·25. Find the total cost of units and rental.

5 From April to July 668 units (at 3p each) are used. If the rental is £8·25, find the total cost of units and rental.

6 In a quarter, 520 units (at 3p each) are dialled and in addition £2·45 is spent on calls via the operator. The rental is £8·25. Find
  a the total cost of all calls (dialled and via operator)
  b the total cost of all calls and rental.

7 If 724 units (at 3p each) are dialled and £0·86 is spent on calls via the operator, find
  a the cost of all the calls made
  b the total cost of calls and rental, if the rental is £8·25.

8 Find the total cost of calls and rental from this information.

Rental for one quarter	£8·25
248 dialled units	3p each
Calls via operator	£1·04
International calls	£5·60

9 VAT is charged on the total cost at a rate of 15%.
Find 15% of
  a £24·60   **b** £32·80   **c** £61·20
  d £17·40   **e** £27·00   **f** £50·80
  g £41·60   **h** £38·80   **i** £44·40
  j £16·20.

Household bills

10 If the VAT is not to an exact penny, then the tax is rounded *down* to the nearest penny. For example, £1·7184 is charged as £1·71.
Calculate 15% of the following total costs rounding down the answers.

a	£24·62	**b**	£32·84	**c**	£41·03
d	£26·12	**e**	£18·95	**f**	£50·12
g	£48·25	**h**	£60·04	**i**	£12·96
j	£8·42				

11 Calculate

a the cost of 364 dialled units (at 3p each)

b the total cost of dialled units and rental at £8·25

c the VAT at 15% on this total

d the amount to be paid (units + rental + VAT).

12

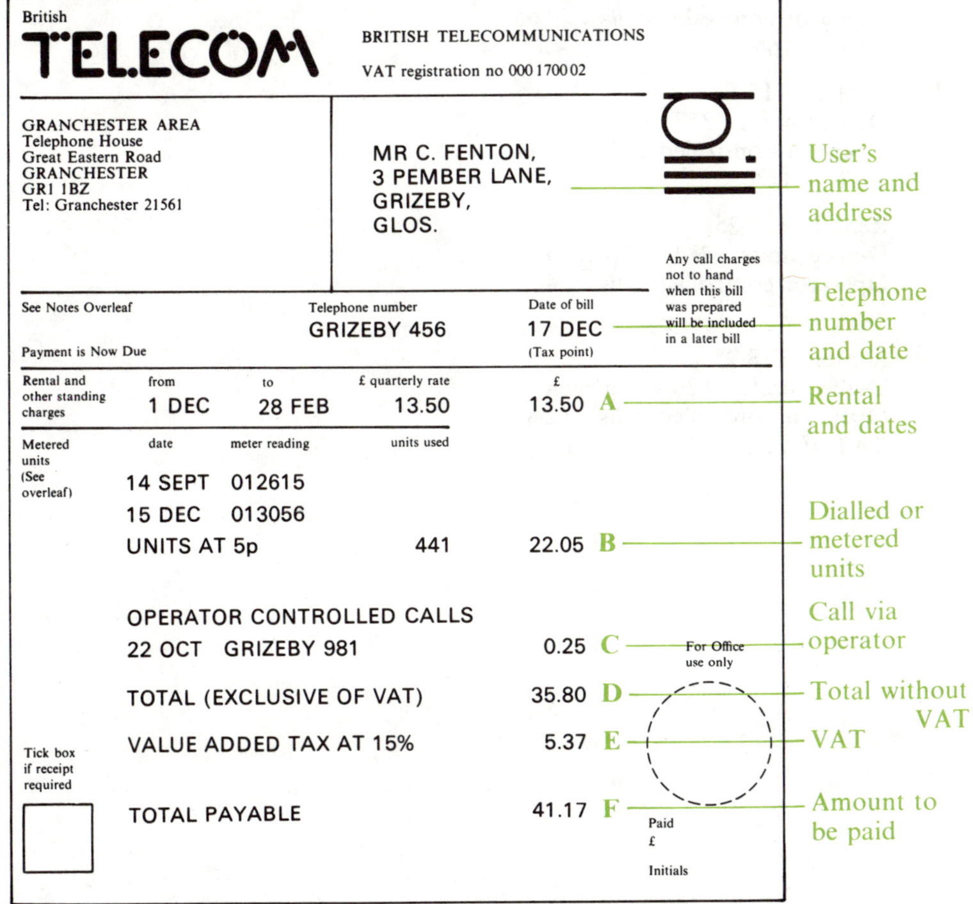

Check the items on the telephone bill shown above, as follows

a check item **B** by subtracting the meter readings to find the number of units used and then work out their cost at 5p each unit

b check item **D** by adding together items **A**, **B** and **C**

c check item **E** by working out 15% of £35·80 rounding down to the nearest penny

d check item **F** by adding items **D** and **E** together.

Household bills

Using the information given, list the items **A** to **F** and make out a bill for
each of these.

13   Rental £13·25
      Dialled units 532 at 4p each
      Operator controlled calls £1·24
      VAT at 15%

14   Rental £13·50
      Dialled units 672 at 4p each
      Operator controlled calls £2·13
      VAT at 12%

15   Rental £15·00
      Dialled units 248 at 4p each
      Operator controlled calls £0·65
      VAT at 16%

16   Rental £15·50
      Dialled units 357 at 5p each
      Operator controlled calls £3·14
      VAT at 14%

17   Rental £17·25
      Dialled units 284 at 5p each
      Operator controlled calls £4·29
      VAT at 13%

18   Rental £18·75
      Dialled units 526 at 6p each
      Operator controlled calls £2·84
      VAT at 12%

Conversions and foreign exchange

Part 1 Using graphs

1 Use this graph to convert pounds sterling (£) into Dutch florins.

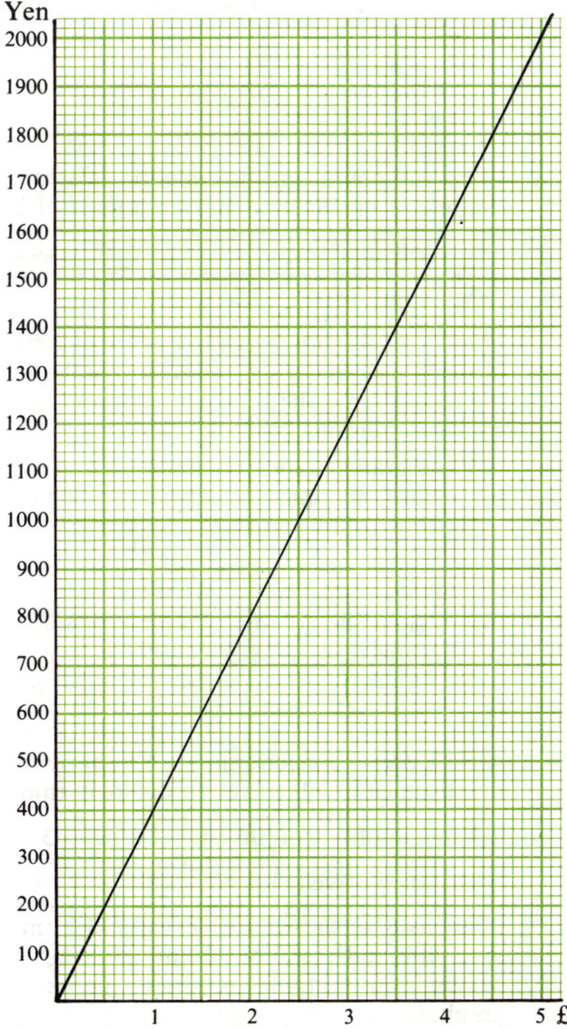

(i) How many florins will you get for

a	£1	b	£3
c	£5	d	£1·40
e	£2·60	f	£3·20
g	£4·40	h	£0·60
i	£3·80	j	£1·60?

(ii) How many pounds £ will you get for

a	20 fl	b	10 fl	c	11 fl	d	12 fl	e	6 fl
f	22 fl	g	18 fl	h	9 fl	i	17 fl	j	23 fl?

2 This conversion graph can be used to change pounds sterling (£) into Japanese yen.

(i) How many yen will you get for

a	£1	b	£3
c	£4	d	£1·50
e	£2·50	f	£4·50
g	£2·25	h	£3·25
i	£4·25	j	£4·75
k	£0·75	l	£2·20
m	£3·20	n	£1·40
o	£2·80?		

(ii) How many pounds £ will you get for

a	800 yen	b	1400 yen
c	200 yen	d	1800 yen
e	1300 yen	f	100 yen
g	700 yen	h	1100 yen
i	1900 yen	j	440 yen
k	520 yen	l	560 yen
m	1240 yen	n	1480 yen
o	1960 yen?		

Conversions and foreign exchange

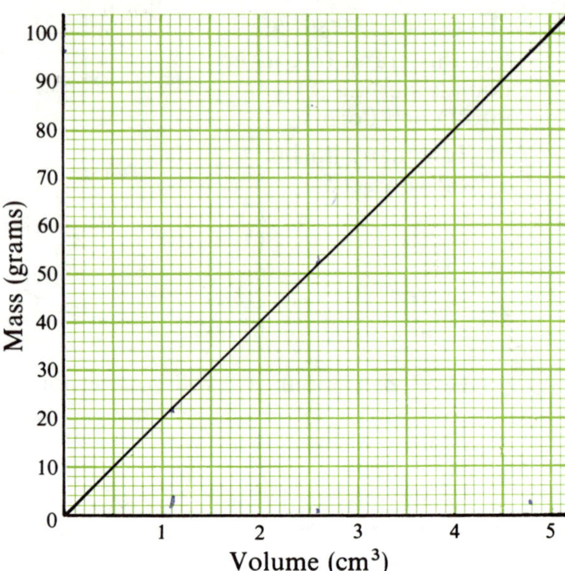

3 This graph gives the mass in grams of any piece of gold with a volume up to 5 cm³.

(i) What is the mass of a piece of gold which has a volume of

a 1 cm³ b 2 cm³
c 5 cm³ d 1.5 cm³
e 2.5 cm³ f 3.2 cm³
g 4.8 cm³ h 2.6 cm³
i 1.1 cm³ j 0.6 cm³?

(ii) What is the volume of a piece of gold which has a mass of

a 60 grams b 80 grams
c 70 grams d 90 grams
e 92 grams f 28 grams
g 36 grams h 84 grams
i 66 grams j 16 grams?

4 Copy and complete this table for different volumes of silicon, taking each cubic centimetre of silicon as having a mass of 2.5 grams.

Volume (cm³)	1	2	3	4
Mass (grams)	2.5			

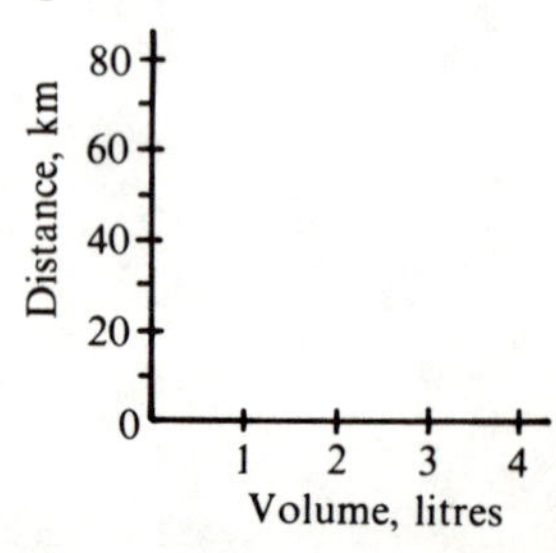

Draw axes as shown (using a scale of 2 cm for each cm³ and 1 cm for each gram), plot the results from the table and draw a conversion graph.

Use the graph to answer these.
(i) What is the mass of a piece of silicon with a volume of

a 2.4 cm³ b 3.6 cm³ c 2.8 cm³
d 0.8 cm³ e 1.8 cm³?

(ii) What is the volume of a piece of silicon with a mass of
a 4 grams b 2 grams c 8 grams d 3.5 grams
e 1.5 grams?

5 A car's petrol consumption is 15 km per litre. Copy and complete this table for the distance it will travel on different amounts of petrol.

Volume of petrol (litres)	1	2	3	4
Distance travelled (km)	15			

Draw axes as shown (using scales of 2 cm for 1 litre and 2 cm for 20 km), plot the results from the table and draw a conversion graph.

Conversions and foreign exchange

Use the graph to answer these.
(i) How far will the car travel on

a 1.2 litres	b 2.6 litres	c 3.4 litres	d 3.8 litres
e 0.8 litres	f 1.6 litres	g 0.6 litres?	

(ii) How many litres of petrol will be needed to travel a distance of

a 21 km	b 36 km	c 42 km	d 48 km
e 54 km	f 27 km	g 33 km?	

6 The best cut of beef costs £5 per kg. Copy and complete this table for different sizes of this cut of beef.

Mass (kg)	1	2	3	4
Cost (£)	5			

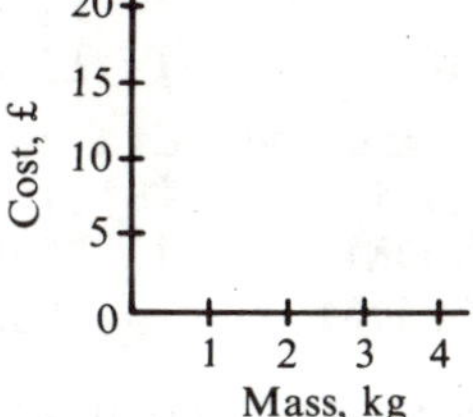

Draw axes as shown, (using scales of 2 cm for 1 kg and 2 cm for £5), plot the results and draw a conversion graph.

Use the graph to answer these.
(i) What is the cost of a piece of beef with a mass of

a 2.2 kg	b 3.2 kg	c 1.4 kg	d 2.4 kg
e 2.6 kg	f 3.8 kg	g 0.8 kg?	

(ii) What is the mass of a piece of beef costing

a £6	b £16	c £17	d £8
e £9	f £14	g £18?	

Part 2 Foreign exchange

1 Take the Anglo-Norwegian rate of exchange as £1 = 8 kroner.
(i) How many kroner will you receive for

a £3	b £4	c £6	d £12
e £18	f £24	g £1·50	h £2·50
i £7·50	j £9·25	k £5·25	l £11·75?

(ii) How many pounds £ will you receive for

a 16 kr.	b 40 kr.	c 64 kr.	d 88 kr.
e 76 kr.	f 84 kr.	g 260 kr.	h 196 kr.
i 186 kr.	j 338 kr.	k 286 kr.	l 542 kr.?

2 Take the Anglo-French rate of exchange as £1 = 12 francs.
(i) How many francs will you receive for

a £2	b £5	c £6	d £17
e £35	f £47	g £7·50	h £12·50
i £9·25	j £24·25	k £42·75	l £51·75?

(ii) How many pounds £ will you receive for

a 48 fr.	b 36 fr.	c 84 fr.	d 288 fr.
e 420 fr.	f 336 fr.	g 18 fr.	h 66 fr.
i 510 fr.	j 747 fr.	k 987 fr.	l 597 fr.?

Conversions and foreign exchange

3 Take the Anglo-Austrian exchange rate as £1 = 32 schillings.
(i) How many schillings will you receive for these amounts?

a	£4	b	£8	c	£23	d	£52
e	£37·50	f	£56·25	g	£41·60	h	£28·40

(ii) How many pounds £ will you receive for these amounts?

a	384 sch.	b	736 sch.	c	720 sch.	d	1008 sch.
e	648 sch.	f	968 sch.	g	3888 sch.	h	4696 sch.

4 When you went on a holiday to Germany the exchange rate was £1 = 3.90 DM.
(i) How many Deutschmarks did you receive for these amounts?

a	£3	b	£8	c	£12	d	£23
e	£148	f	£2·60	g	£7·40	h	£58·30
i	£92·70	j	£306·40				

(ii) How many pounds £ did you receive for these amounts?

a	7.80 DM	b	11.70 DM	c	19.50 DM	d	31.20 DM
e	46.80 DM	f	81.90 DM	g	83.07 DM	h	60.06 DM
i	134.94 DM	j	203.97 DM				

5 When you went on a holiday to the United States the rate of exchange was £1 = $1.80.
(i) How many dollars did you receive for these amounts?

a	£4	b	£7	c	£24	d	£132
e	£265	f	£6·35	g	£4·85	h	£58·70
i	£46·85	j	£362·75				

(ii) How many pounds £ did you receive for these amounts?

a	$5.40	b	$3.78	c	$2.34	d	$22.50
e	$61.56	f	$36.81	g	$112.23	h	$261.54
i	$456.48	j	$944.37				

6 Take the Anglo-Canadian exchange rate as £1 = $2.10.
(i) How many dollars will you receive for these amounts? Give the answers correct to the nearest cent (where $1 = 100 cents).

a	£3·52	b	£6·54	c	£8·27	d	£12·72
e	£30·26	f	£125·70	g	£204·66	h	£650·78

(ii) How many pounds £ will you receive for these Canadian dollars? Give the answers to the nearest penny.

a	$6.56	b	$5.11	c	$26.34	d	$49.22
e	$67.25	f	$50.00	g	$120.00	h	$300.00

7 While on holiday in France, where the exchange rate is £1 = 11 francs, you go to a bank and change £65 into francs. How many francs do you get?

8 Amanda, who is on holiday in Italy, is willing to spend up to £8 on a present for her mother. If £1 = 1760 lire, what is the maximum number of lire she would spend?

9 The airport tax at a Spanish airport is £4·50 when paid in British currency. If £1 = 184 pesetas, how much is this tax in pesetas?

10 In Britain, a book costs £9·45. When visiting relatives in Australia, Martin sees the same book priced at $15.44. If £1 = $1.60, which country has the lower price for the book?

11 If I can get 4 Deutschmarks for £1, how much in British currency is the cost of a 24-DM boat-trip on the river Rhine?

12 A pair of trousers in a Danish shop cost 175 kr. If £1 = 14 kroner, find their cost in British currency.

Conversions and foreign exchange

13 Whilst buying presents at the end of a holiday in Switzerland, Julie sees a box of Swiss chocolates at 8.75 francs. If £1 = 3.50 francs, how much do they cost in British money?

14 A packet of Dutch cigars cost 12.25 florins in Holland where £1 can be changed for 5 florins. How much do they cost in pounds sterling?

15 I get 2.20 Canadian dollars for each £1. If a present costs me $7.15, how much is this in pounds sterling?

16 I can change £120 into French francs at a rate of £1 = 10.5 francs. If I spend 987 francs, how much have I left
    a in francs   **b** in pounds £, if the exchange rate has not changed?

Part 3 Flow diagrams

1 This flow diagram can be used to change a mass measured in kilograms (kg) to a mass measured in pounds (lb).

Change these masses into pounds (lb).
a	10 kg	**b**	20 kg	**c**	35 kg	**d**	15 kg
e	40 kg	**f**	60 kg	**g**	75 kg	**h**	85 kg
i	135 kg						

Use the flow diagram *in reverse* to change these masses into kilograms (kg).
j	55 lb	**k**	66 lb	**l**	110 lb	**m**	99 lb
n	132 lb	**o**	275 lb	**p**	385 lb	**q**	495 lb
r	715 lb						

2 Volumes in gallons can be changed into litres using this flow diagram.

Change these volumes into litres.
a	10 gallons	**b**	12 gallons	**c**	24 gallons	**d**	32 gallons
e	46 gallons	**f**	56 gallons	**g**	5 gallons	**h**	11 gallons
i	13 gallons						

Use the flow diagram *in reverse* to change these volumes into gallons.
j	27 litres	**k**	36 litres	**l**	72 litres	**m**	99 litres
n	306 litres	**o**	288 litres	**p**	369 litres	**q**	549 litres
r	738 litres						

3 Pounds sterling (£) can be changed into Australian dollars ($) using this flow diagram.

Change these amounts into dollars.
a	£20	**b**	£35	**c**	£15	**d**	£40
e	£55	**f**	£95	**g**	£125	**h**	£165
i	£235						

Use the flow diagram *in reverse* to change these amounts into £.
j	$16	**k**	$64	**l**	$72	**m**	$96
n	$136	**o**	$144	**p**	$184	**q**	$224
r	$376						

Conversions and foreign exchange

4 This flow diagram can be used to change temperatures in °C to those in °F.

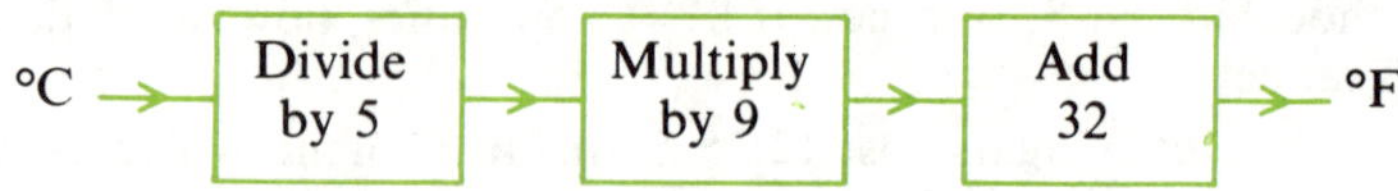

Change these temperatures into °F.

a 20°C b 15°C c 35°C d 40°C
e 0°C f 100°C g 75°C h −10°C
i −20°C

Using the flow diagram *in reverse*, change these temperatures into °C.

j 50°F k 77°F l 41°F m 86°F
n 122°F o 113°F p 185°F q 23°F
r 5°F

5 This flow diagram gives inland postal rates for parcels with masses up to 5 kg. A lower Area Rate applies for parcels delivered locally; outside this area, the National Rate applies.

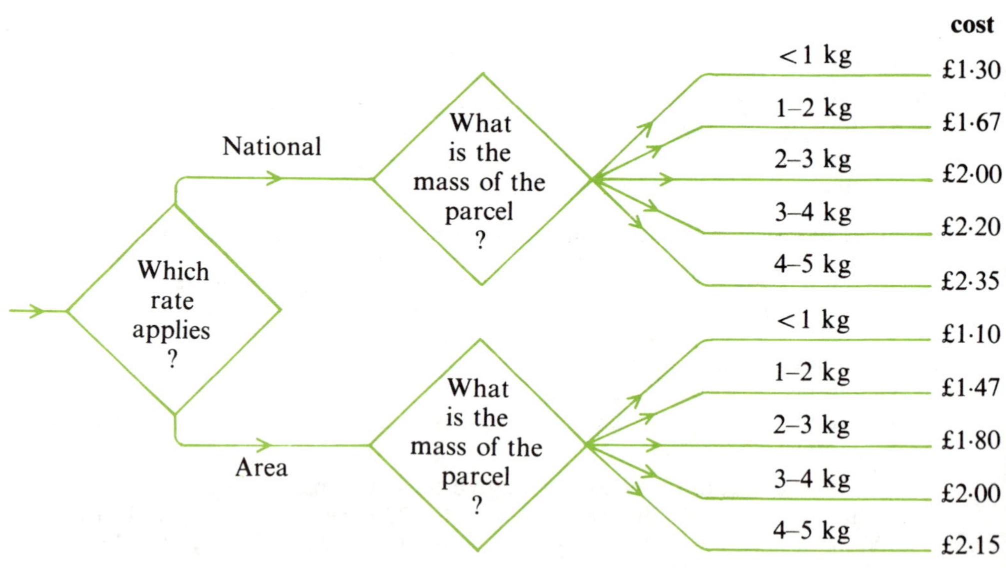

How much will it cost to send these parcels?

a a $2\frac{1}{2}$-kg parcel locally
b a $3\frac{1}{2}$-kg parcel nationally
c a 3.25-kg parcel nationally
d two 4.4-kg parcels locally
e five 1.25-kg parcels nationally
f three $2\frac{1}{4}$-kg and four $\frac{3}{4}$-kg parcels, all locally
g six $4\frac{1}{2}$-kg and eight $3\frac{3}{4}$-kg parcels nationally
h nine $\frac{1}{2}$-kg parcels locally and twelve $1\frac{1}{4}$-kg ones nationally
i eight 2.7-kg parcels nationally and two 0.8-kg parcels locally
j four 4.2-kg parcels locally, five 3.7-kg parcels nationally and fourteen 0.65-kg ones locally

Conversions and foreign exchange

6 This flow diagram gives inland postal rates for letters upto a mass of 200 grams. Both first and second class rates are given.

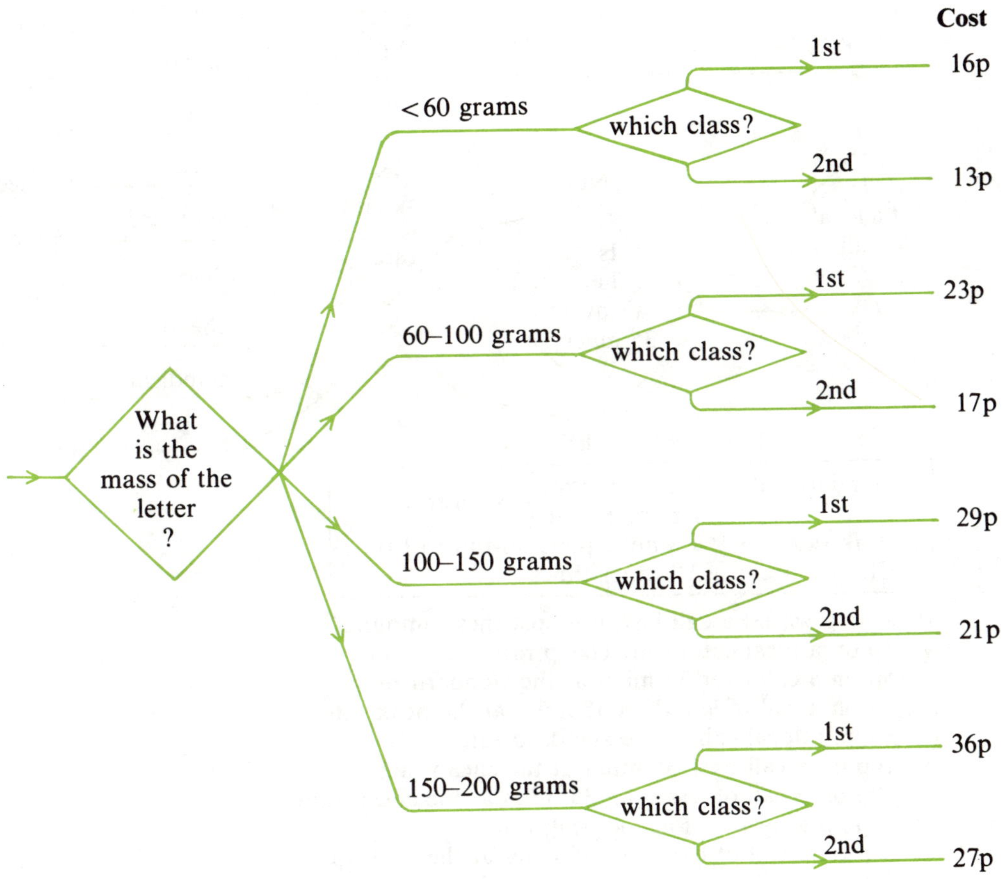

How much will it cost to send these letters?
a a 45-gram first-class letter
b a 75-gram first-class letter
c a 120-gram second-class letter
d a 185-gram second-class letter
e two 50-gram letters, one of each class
f four 95-gram letters, two of each class
g eight 110-gram first-class letters
h twelve Christmas cards, all less than 60 grams and all first-class
i six 2nd-class birthday cards of 90 grams each
j fifteen 125-gram first-class letters and sixteen 175-gram second-class letters

Conversions and foreign exchange

7 This flow diagram gives telephone charges for calls dialled direct. The diagram gives the length of time allowed for 5 pence (including VAT).

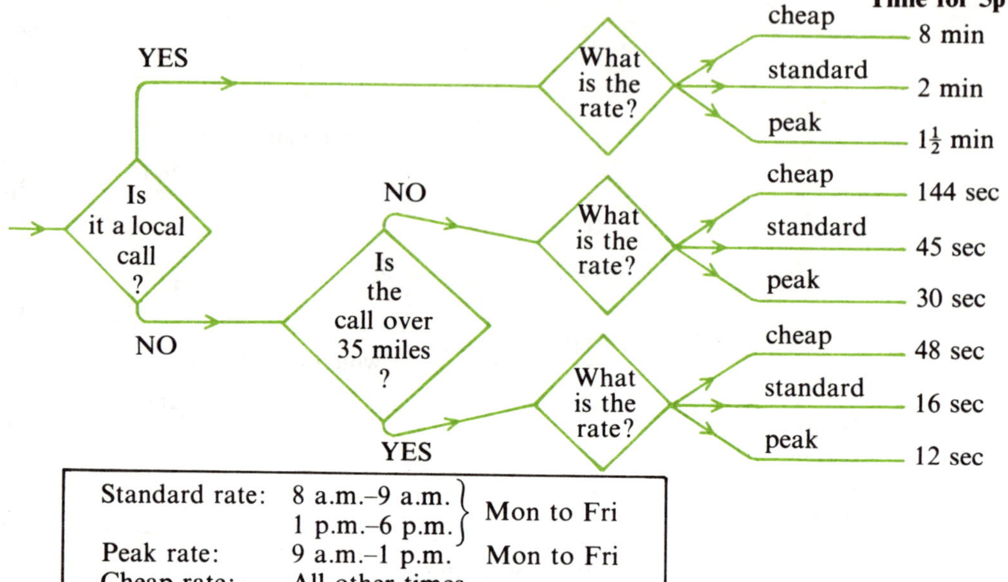

Standard rate:	8 a.m.–9 a.m. } Mon to Fri
	1 p.m.–6 p.m. }
Peak rate:	9 a.m.–1 p.m. Mon to Fri
Cheap rate:	All other times

How long would a call last, if it cost these amounts?
a 5p on a local call at the cheap rate
b 5p on a call over 35 miles at the standard rate
c 5p on a call of less than 35 miles at the peak rate
d 10 on a local call at the standard rate
e 10p on a call over 35 miles at the cheap rate
f 10p on a call of less than 35 miles at the cheap rate
g 20p on a local call at the peak rate
h 25p on a call of less than 35 miles at the peak rate
i 30p on a call over 35 miles at the cheap rate
j 45p on a local call at the standard rate
k 50p on a call of less than 35 miles at the standard rate
l 60p on a call over 35 miles at the peak rate

How much would it cost to make these telephone calls?
m 8 minutes locally at the cheap rate
n 16 minutes locally at the cheap rate
o 30 seconds of less than 35 miles at the peak rate
p 1 minute of less than 35 miles at the peak rate
q 12 seconds of more than 35 miles at the peak rate
r 1 minute of more than 35 miles at the peak rate
s 10 minutes locally at the standard rate
t 4 minutes of less than 35 miles at the peak rate
u 6 minutes locally at the peak rate
v 1½ minutes of less than 35 miles at the standard rate
w 48 seconds of more than 35 miles at 10 a.m. on a Tuesday morning
x 24 minutes locally at 2 p.m. on a Saturday afternoon
y 3½ minutes of less than 35 miles at 11 a.m. on a Friday morning
z 6 minutes of less than 35 miles at 3.30 p.m. on a Tuesday afternoon

The 24-hour clock and timetables

1 Change these times
into those on the 24-hour clock.
a   1.30 p.m.   b   2.15 p.m.
c   4.25 p.m.   d   8.45 p.m.
e   12.40 p.m.   f   8.10 a.m.
g   7.25 a.m.   h   10.30 a.m.
i   11.55 a.m.   j   11.55 p.m.
k   9.45 a.m.   l   9.45 p.m.
m   ten to eight in the morning
n   a quarter past seven in the evening
o   half past ten at night
p   a quarter to midnight
q   a quarter past midnight
r   ten past two in the early morning
s   twenty-five to four in the afternoon
t   fourteen minutes to ten in the morning

2 These times are written for the 24-hour clock.
Rewrite them using a.m. and p.m.
a   13.20   b   17.25   c   20.35   d   22.25   e   09.10
f   07.40   g   08.55   h   10.35   i   11.00   j   12.42
k   01.23   l   02.57   m   00.17   n   12.17   o   00.54
p   12.54

3 Which of these statements are *true* and which are *false*?
a   five past five in the afternoon = 17.05
b   quarter to eleven in the morning = 11.45
c   half past six in the morning = 06.30
d   five to nine at night = 21.55
e   twenty past eleven in the morning = 11.20
f   ten to eight in the evening = 19.50
g   ten to midnight = 23.10
h   2.15 p.m. = 02.15
i   6.42 a.m. = 06.42
j   8.54 p.m. = 20.54
k   3 o'clock in the morning = 15.00
l   five to six in the evening = 18.05
m   twenty past midnight = 12.20

Give the answers to the following problems on the 24-hour clock.

4 a   A man sets out for work at 07.50 and the journey takes him 45 minutes.
At what time does he get to work?
b   The return journey home starts at 17.10 and takes him 1 hour 5 minutes.
When does he arrive home?

5 It takes $2\frac{3}{4}$ hours to drive to the coast. What time do you arrive there, if you set
out at
a   08.15   b   11.30   c   12.05?

6 John Bradley takes 70 minutes to cycle to the city centre. At what time does he
arrive there, if he sets out at
a   09.15   b   13.25   c   14.55?

The 24-hour clock and timetables

7 All trains from Saxton to Angleby take $1\frac{1}{4}$ hours. At what times do they arrive in Angleby, if they leave Saxton at
   a 10.35   **b** 17.20   **c** 19.50?

8 A hill-walker leaves a Youth Hostel at 09.25 to walk 15 miles.
   a If he expects to average 2 miles every hour, at what time will he finish his hike?
   b If he stops for $\frac{3}{4}$ hour on the way, when will his journey be over?

9 Bristol to Kendal by motorway is 225 miles. If a car averages 50 miles in every hour, when will it arrive in Kendal, if it leaves Bristol at
   a 06.15   **b** 17.45   **c** 22.35?

10 Mrs Williams leaves home at 09.25 to go to London. She takes 50 minutes to get to the railway station where she had to wait 20 minutes before catching the train which takes $2\frac{1}{4}$ hours to reach London. At what time does she arrive?

11 Three Inter-City overnight expresses each take $6\frac{3}{4}$ hours for their journey. What time do they arrive, if they leave at
   a 21.05   **b** 23.15   **c** 23.50?

12 Jim Watson intends cycling 100 kilometres in a day. If he leaves home at 07.50, when will he finish his trip, if he averages a speed of
   a 10 kilometres per hour   **b** 12 kilometres per hour
   c 16 kilometres per hour?

13 An aeroplane takes 1 hour 25 minutes for a journey starting at 21.40.
   a What is the expected time of arrival?
   b If the plane is $2\frac{1}{4}$ hours late taking-off because of fog, what will be the time of arrival?

14 Mr Baldwin leaves home at 08.20 and takes $1\frac{1}{4}$ hours to reach the airport where he has a $1\frac{1}{2}$-hour wait before take-off. The flight lasts 3 hours 50 min.
   a At what time does he land?
   b If the flight is 25 minutes late, at what time does he land?

15 (i) A coach journey takes 5 h 20 min. What is the time of arrival at the destination if the departure time is
   a 18.30   **b** 18.40   **c** 18.50   **d** 19.50
   e 20.10   **f** 22.30   **g** 23.40   **h** 23.55?

(ii) A train journey takes 7 h 35 min. What is the time of arrival at the destination if the departure time is
   a 16.20   **b** 16.25   **c** 16.30   **d** 20.15
   e 21.25   **f** 22.40   **g** 23.10   **h** 23.55?

16 Penzance to Carlisle is 460 miles.
   a If a car leaves Penzance at 16.35 and averages a speed of 60 miles per hour, at what time will it arrive in Carlisle?
   b If traffic conditions reduce the average speed to 50 miles per hour, what will be the arrival time then?

17 A cross-channel ferry leaves Calais at 21.40 and takes $3\frac{1}{4}$ hours to reach Dover. There is a 20-minute wait before the London train leaves Dover, and the train arrives in London 2 hours 35 min later. Find the time of arrival in London.

The 24-hour clock and timetables

18 This table gives times of buses in both directions between Barnsley and Manchester.

Barnsley - Manchester via Penistone and Holmfirth
⚆ Adjoining or near Railway Station. **S** Saturday only.

Daily		S		
Barnsley Bus Station ⚆	09.05	12.05	15.05	18.05
Dodworth Crossroads	09.14	12.14	15.14	18.14
Silkstone Common	09.16	12.16	15.16	18.16
Oxspring Waggon & Horses	09.22	12.22	15.22	18.22
Penistone Church ⚆	09.27	12.27	15.27	18.27
Sovereign Inn	09.39	12.39	15.39	18.39
New Mill Crossroads	09.45	12.45	15.45	18.45
Holmfirth Victoria Square	09.51	12.51	15.51	18.51
Greenfield Clarence Hotel	10.11	13.11	16.11	19.11
Mossley Station ⚆	10.15	13.15	16.15	19.15
Stalybridge Thompson Cross	10.21	13.21	16.21	19.21
Ashton-under-Lyne Bus Stn ⚆	10.25	13.25	16.25	19.25
Manchester Chorlton St ⚆	10.50	13.50	16.50	19.50

Daily		S		
Manchester Chorlton St ⚆	11.05	14.05	17.05	20.05
Ashton-under-Lyne Bus Stn ⚆	11.30	14.30	17.30	20.30
Stalybridge Thompson Cross	11.33	14.33	17.33	20.33
Mossley Station ⚆	11.36	14.36	17.36	20.36
Greenfield Clarence Hotel	11.42	14.42	17.42	20.42
Holmfirth Bus Station	12.04	15.04	18.04	21.04
New Mill Crossroads	12.10	15.10	18.10	21.10
Sovereign Inn	12.16	15.16	18.16	21.16
Penistone Church ⚆	12.28	15.28	18.28	21.28
Oxspring Waggon & Horses	12.33	15.33	18.33	21.33
Silkstone Common	12.39	15.39	18.39	21.39
Dodworth Crossroads	12.41	15.41	18.41	21.41
Barnsley Bus Station ⚆	12.50	15.50	18.50	21.50

a What does the letter *S* stand for above the 15.05 from Barnsley?

b How many buses leave Barnsley for Manchester after 12 o'clock midday on any weekday?

c If you want to get to Manchester by 3 p.m. on a Friday, what time would you have to leave Barnsley?

d You are in Manchester and want to arrive in Barnsley before 9 p.m. on a Sunday. At what time would the most convenient bus leave Manchester?

e How long does the whole journey take?

f You live in Penistone and want to get to Holmfirth before 1 p.m. on a Saturday. At what time would you catch a bus in Penistone, and how long would your journey take?

g You live in Stalybridge and want to get to New Mill by 4 p.m. on a Wednesday. What time will the bus leave Stalybridge and how long will your journey take?

h You live in Dodworth and you want to visit your Auntie Mary who lives in Ashton-under-Lyne, travelling there and back on the same day. You leave on the first possible bus and return on the latest possible bus.
   (i) When do you leave Dodworth?
   (ii) When do you arrive in Ashton?
   (iii) When do you leave Ashton on the return trip?
   (iv) When do you arrive back in Dodworth?
   (v) How long have you spent travelling on a bus during this day?

19 On page 182 is the British Rail timetable for the York–Sheffield weekday service. Trains stop at up to six stations on this route.

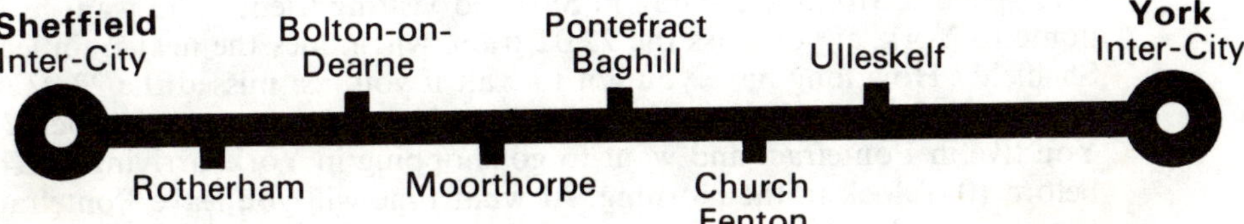

The 24-hour clock and timetables

York - Sheffield

Mondays to Saturdays

York *P*	d.	07.18	08.20	08.16	09.26	10.18	10.30	11.04	11.30	11.59	12.37	13.20	13.44	15.24	16.32
Ulleskelf*	d.										12.49				16.44
Church Fenton *FP*	d.	07.32									12.53				16.48
Pontefract Baghill *FP*	d.	07.46				10.44	10.59	11.30			13.07				17.02
Moorthorpe *FP*	d.	07.55				10.53	u11.10	11.41			13.17				17.12
Bolton-on-Dearne *FP*	d.	08.07				11.05	u11.25	11.55			13.28				17.23
Rotherham *P*	d.	08.19		09.38		11.17	u11.38	12.08			13.41				17.36
Sheffield *P*	a.	08.31	09.20	09.51	10.26	11.29	11.51	12.21	12.30	13.40	13.52	14.21	14.45	16.26	17.47

York *P*	d.	17.29	17.34	17.52	19.23	19.38	21.34
Ulleskelf*	d.					19.52	
Church Fenton *FP*	d.						
Pontefract Baghill *FP*	d.		18.00		19.49	20.08	22.05
Moorthorpe *FP*	d.		18.09			20.18	
Bolton-on-Dearne *FP*	d.		18.21			20.29	
Rotherham *P*	d.		18.34		20.21	20.43	
Sheffield *P*	a.	18.29	18.45	18.52	20.35	20.54	22.46

Notes:
a Arrive. d Depart.
u Stops to pick up only.
* No staff in attendance.
P Car parking available-fee payable.
FP Free car parking at or near station.

Sheffield - York

Mondays to Saturdays

Sheffield *P*	d.	00.50	07.05	08.22	08.55	09.02	10.00	10.25	10.53	11.09	12.42	12.50	13.00	13.12	13.17
Rotherham *P*	d.		07.15	08.32	09.04			10.35				12.59			13.27
Bolton-on-Dearne *FP*	d.		07.27	08.44	09.19			10.47							13.39
Moorthorpe *FP*	d.		07.36	08.53	09.29			10.56							13.48
Pontefract Baghill *FP*	d.		07.47	09.04	09.40			11.07							13.59
Church Fenton *FP*	d.														
Ulleskelf*	d.		08.06		09.59										14.18
York *P*	a.	01.52	08.21	09.35	10.12	10.01	11.00	11.38	11.54	12.08	13.43	13.54	13.59	14.12	14.35

Sheffield *P*	d.	14.40	15.06	15.12	15.20	16.25	17.17	17.42	17.45	18.01	19.28	20.18		22.02
Rotherham *P*	d.					16.35	17.25		17.55	18.10				22.12
Bolton-on-Dearne *FP*	d.	15.00				16.47			18.07					22.24
Moorthorpe *FP*	d.	15.12				16.56			18.16					22.33
Pontefract Baghill *FP*	d.	15.23				17.07			18.27	18.38			21.55	22.44
Church Fenton *FP*	d.					17.24								
Ulleskelf*	d.													
York *P*	a.	15.56	16.05	16.13	16.19	17.40	18.23	18.43	18.58	19.11	20.29	21.17	22.24	23.19

a What do the letters *a* and *d* stand for after the names of the stations?

b What does the sign * stand for printed after the station name Ulleskelf?

c What do the letters *FP* stand for?

d How many trains leave York for Sheffield before 12 midday?

e How many trains stop at Church Fenton on their way from York to Sheffield?

f How many trains travel each day direct from Sheffield to York without stopping on the way?

g How long does it take to get to York
  (i) on the 10.00 train from Sheffield
  (ii) on the 10.53 train from Sheffield
  (iii) on the 11.09 train from Sheffield?

h How many stations does the 17.45 train from Sheffield stop at before reaching York, and how long does it take for the whole journey?

i You live in York and have to visit Sheffield, arriving there by 11 a.m. At what time does the latest possible train leave York?

j You spend a Monday evening in Sheffield visiting friends and want to go home to York. If you miss the 22.02 train, when does the next train leave Sheffield? How long have you got to wait if you *just* missed the 22.02 train?

k How long does the 10.44 train from Pontefract take to reach Sheffield?

l You live in Pontefract and want to go shopping in York, arriving in York before 10 o'clock in the morning. At what time will you leave Pontefract and how long will the journey take you?

The 24-hour clock and timetables

m Shops in York close at 5 p.m. If you return to Pontefract as soon as possible after they close, what time would you leave York and what time would you arrive back in Pontefract?

n A workman, who lives in Pontefract, finishes work in Bolton-on-Dearne at 5.30 p.m. What time will his train leave Bolton-on-Deane, and how long will it take to get to Pontefract?

20 This bus timetable gives the service from Huddersfield to Bradford throughout the week.

Huddersfield - Bradford

Monday to Friday

Huddersfield Bus Stn	05.25	Then at	40	55	10	25		17.40	18.00	18.20	18.35	18.50	19.05
Brighouse	05.45	these	00	15	30	45	until	18.00	18.20	18.40	18.55	19.10	19.25
Bailiff Bridge	05.50	min past	05	20	35	50		18.05	18.25	18.45	19.00	19.15	19.30
Bradford The Tyrls	06.10	each hour	25	40	55	10		18.25	18.45	19.05	19.20	19.35	19.50

Huddersfield Bus Stn	Then at	25	55		22.25	22.55
Brighouse	these	45	15	until	22.45	23.15
Bailiff Bridge	min past	50	20		22.50	23.20
Bradford The Tyrls	each hour	10	40		23.10	23.40

Saturday

Huddersfield Bus Stn	05.25	05.55	06.25	06.55	07.25	Then at	40	55	10	25		22.55	23.10
Brighouse	05.45	06.15	06.45	07.15	07.45	these	00	15	30	45		23.15	23.30
Bailiff Bridge	05.50	06.20	06.50	07.20	07.50	min past	05	20	35	50	until	23.20	23.35
Bradford The Tyrls	06.10	06.40	07.10	07.40	08.10	each hour	25	40	55	10		23.40	23.55

Sunday

Huddersfield Bus Stn	09.25	Then at	55	25		22.55
Brighouse	09.45	these	15	45	until	23.15
Bailiff Bridge	09.50	min past	20	50		23.20
Bradford The Tyrls	10.10	each hour	40	10		23.40

a How long does it take to go from
(i) Huddersfield to Bradford   (ii) Huddersfield to Bailiff Bridge
(iii) Brighouse to Bradford?

b When does the first bus leave Huddersfield for Bradford on
(i) Wednesday   (ii) Saturday   (iii) Sunday?

c When does the last bus leave Brighouse for Bradford on
(i) Wednesday   (ii) Saturday   (iii) Sunday?

d If on Monday you miss the 05.25 bus from Huddersfield to Bradford, when does the next bus leave Huddersfield?

e If on Monday you miss the 12.10 bus from Huddersfield to Bradford, when does the next bus leave Huddersfield?

f If on Monday you miss the 17.25 bus out of Huddersfield, at what time does the next bus leave Huddersfield?

g How many minutes are there between buses on this route, before 6 p.m. Monday to Friday?

h How often do the buses run on this route on a Sunday?

i You live in Huddersfield and you want to get to Bradford by 7 p.m. on Monday. What is the latest possible time you can catch a bus from Huddersfield?

j You live in Brighouse and you want to get to Bradford by 4 p.m. on Saturday. What is the latest possible time you can catch a bus from Brighouse?

k On Saturday, you just miss the 20.55 bus from Huddersfield to Bailiff Bridge. When will the next bus leave Huddersfield dand when will it arrive in Bailiff Bridge?

The 24-hour clock and timetables

21   A bus service visits the towns on this timetable. Copy the timetable and complete it by filling in the missing times, using the information given here.

Cardiff	08.40		
Newport		12.35	
Monmouth			20.10
Gloucester			
Cheltenham			

The journey from
    Cardiff to Newport takes 25 minutes
    Newport to Monmouth takes 40 minutes
    Monmouth to Gloucester takes 45 minutes
    Gloucester to Cheltenham takes 15 minutes.

22   Copy this timetable and fill in the missing times, using the information given.

The journey from
    Penzance to Cambourne takes 40 minutes
    Penzance to Redruth takes 50 minutes
    Penzance to Truro takes $1\frac{1}{4}$ hours
    Penzance to Bodmin takes 1 hour 55 minutes
    Penzance to Liskeard takes $2\frac{1}{4}$ hours
    Penzance to Plymouth takes 2 hours 50 minutes.

Penzance	07.45	11.50		
Cambourne			14.55	
Redruth				
Truro				20.15
Bodmin				
Liskeard				
Plymouth				

Conversions and Constants

Length

$$10 \text{ mm} = 1 \text{ cm}$$
$$100 \text{ cm} = 1 \text{ m}$$
$$1000 \text{ m} = 1 \text{ km}$$

Area

$$100 \text{ mm}^2 = 1 \text{ cm}^2$$
$$10\,000 \text{ cm}^2 = 1 \text{ m}^2$$
$$10\,000 \text{ m}^2 = 1 \text{ hectare (ha)}$$
$$100 \text{ ha} = 1 \text{ km}^2$$

Volume/capacity

$$1000 \text{ cm}^3 = 1 \text{ litre}$$
$$1000 \text{ litres} = 1 \text{ m}^3$$

Mass

$$1000 \text{ grams} = 1 \text{ kg}$$
$$1000 \text{ kg} = 1 \text{ tonne}$$

Imperial–metric conversions

Length

1 inch	$= 2.54$ cm	1 cm $= 0.394$ inches
1 foot ($= 12$ inches)	$= 30.5$ cm	
1 yard ($= 3$ feet)	$= 0.914$ m	1 m $= 1.09$ yards
1 mile ($= 1760$ yards)	$= 1.61$ km	1 km $= 0.621$ mile
		8 km $= 5$ miles

Area

1 sq. inch	$= 6.45$ cm^2	1 cm^2 $= 0.155$ sq. inch
1 sq. foot	$= 929$ cm^2	
1 sq. yard	$= 0.836$ m^2	1 m^2 $= 1.20$ sq. yards
1 acre ($= 4840$ yd^2)	$= 0.405$ ha	1 ha $= 2.47$ acres
1 sq. mile ($= 640$ acres)	$= 2.59$ km^2	1 km^2 $= 0.386$ sq. mile

Volume/capacity

1 cubic inch	$= 16.4$ cm^3	1 cm^3 $= 0.061$ cu. inch
1 cubic yard	$= 0.765$ m^3	1 m^3 $= 1.31$ cu. yards
1 pint	$= 0.568$ litres	1 litre $= 1.76$ pints
1 gallon ($= 8$ pints)	$= 4.55$ litres	1 litre $= 0.22$ gallon

Mass

1 ounce	$= 28.35$ grams	1 gram $= 0.0353$ ounce
1 pound ($= 16$ ounces)	$= 0.454$ kg	1 kg $= 2.20$ pounds
1 ton ($= 2240$ pounds)	$= 1.02$ tonnes	1 tonne $= 0.984$ ton

The metric prefixes

deca	da	10	deci	d	10^{-1}
hecto	h	10^2	centi	c	10^{-2}
kilo	k	10^3	milli	m	10^{-3}
mega	M	10^6	micro	μ	10^{-6}
giga	G	10^9	nano	n	10^{-9}
tera	T	10^{12}	pico	p	10^{-12}

Constants (to 3 significant figures)

$$\pi = 3.14$$
$$\log \pi = 0.497$$
$$1 \text{ radian} = 57.3°$$

Radius of the Earth $= 6370$ km

Acceleration due to gravity $= 9.81$ m/s^2

Velocity of light $= 3 \times 10^8$ m/s

Logarithms

	0	1	2	3	4	5	6	7	8	9
1.0	0.000	004	009	013	017	021	025	029	033	037
1.1	0.041	045	049	053	057	061	064	068	072	076
1.2	0.079	083	086	090	093	097	100	104	107	111
1.3	0.114	117	121	124	127	130	134	137	140	143
1.4	0.146	149	152	155	158	161	164	167	170	173
1.5	0.176	179	182	185	188	190	193	196	199	201
1.6	0.204	207	210	212	215	217	220	223	225	228
1.7	0.230	233	236	238	241	243	246	248	250	253
1.8	0.255	258	260	262	265	267	270	272	274	276
1.9	0.279	281	283	286	288	290	292	294	297	299
2.0	0.301	303	305	307	310	312	314	316	318	320
2.1	0.322	324	326	328	330	332	334	336	338	340
2.2	0.342	344	346	348	350	352	354	356	358	360
2.3	0.362	364	365	367	369	371	373	375	377	378
2.4	0.380	382	384	386	387	389	391	393	394	396
2.5	0.398	400	401	403	405	407	408	410	412	413
2.6	0.415	417	418	420	422	423	425	427	428	430
2.7	0.431	433	435	436	438	439	441	442	444	446
2.8	0.447	449	450	452	453	455	456	458	459	461
2.9	0.462	464	465	467	468	470	471	473	474	476
3.0	0.477	479	480	481	483	484	486	487	489	490
3.1	0.491	493	494	496	497	498	500	501	502	504
3.2	0.505	507	508	509	511	512	513	515	516	517
3.3	0.519	520	521	522	524	525	526	528	529	530
3.4	0.531	533	534	535	537	538	539	540	542	543
3.5	0.544	545	547	548	549	550	551	553	554	555
3.6	0.556	558	559	560	561	562	563	565	566	567
3.7	0.568	569	571	572	573	574	575	576	577	579
3.8	0.580	581	582	583	584	585	587	588	589	590
3.9	0.591	592	593	594	595	597	598	599	600	601
4.0	0.602	603	604	605	606	607	609	610	611	612
4.1	0.613	614	615	616	617	618	619	620	621	622
4.2	0.623	624	625	626	627	628	629	630	631	632
4.3	0.633	634	635	636	637	638	639	640	641	642
4.4	0.643	644	645	646	647	648	649	650	651	652
4.5	0.653	654	655	656	657	658	659	660	661	662
4.6	0.663	664	665	666	667	667	668	669	670	671
4.7	0.672	673	674	675	676	677	678	679	679	680
4.8	0.681	682	683	684	685	686	687	688	688	689
4.9	0.690	691	692	693	694	695	695	696	697	698
5.0	0.699	700	701	702	702	703	704	705	706	707
5.1	0.708	708	709	710	711	712	713	713	714	715
5.2	0.716	717	718	719	719	720	721	722	723	723
5.3	0.724	725	726	727	728	728	729	730	731	732
5.4	0.732	733	734	735	736	736	737	738	739	740

	0	1	2	3	4	5	6	7	8	9
5.5	0.740	741	742	743	744	744	745	746	747	747
5.6	0.748	749	750	751	751	752	753	754	754	755
5.7	0.756	757	757	758	759	760	760	761	762	763
5.8	0.763	764	765	766	766	767	768	769	769	770
5.9	0.771	772	772	773	774	775	775	776	777	777
6.0	0.778	779	780	780	781	782	782	783	784	785
6.1	0.785	786	787	787	788	789	790	790	791	792
6.2	0.792	793	794	794	795	796	797	797	798	799
6.3	0.799	800	801	801	802	803	803	804	805	806
6.4	0.806	807	808	808	809	810	810	811	812	812
6.5	0.813	814	814	815	816	816	817	818	818	819
6.6	0.820	820	821	822	822	823	823	824	825	825
6.7	0.826	827	827	828	829	829	830	831	831	832
6.8	0.833	833	834	834	835	836	836	837	838	838
6.9	0.839	839	840	841	841	842	843	843	844	844
7.0	0.845	846	846	847	848	848	849	849	850	851
7.1	0.851	852	852	853	854	854	855	856	856	857
7.2	0.857	858	859	859	860	860	861	862	862	863
7.3	0.863	864	865	865	866	866	867	867	868	869
7.4	0.869	870	870	871	872	872	873	873	874	874
7.5	0.875	876	876	877	877	878	879	879	880	880
7.6	0.881	881	882	883	883	884	884	885	885	886
7.7	0.886	887	888	888	889	889	890	890	891	892
7.8	0.892	893	893	894	894	895	895	896	897	897
7.9	0.898	898	899	899	900	900	901	901	902	903
8.0	0.903	904	904	905	905	906	906	907	907	908
8.1	0.908	909	910	910	911	911	912	912	913	913
8.2	0.914	914	915	915	916	916	917	918	918	919
8.3	0.919	920	920	921	921	922	922	923	923	924
8.4	0.924	925	925	926	926	927	927	928	928	929
8.5	0.929	930	930	931	931	932	932	933	933	934
8.6	0.934	935	936	936	937	937	938	938	939	939
8.7	0.940	940	941	941	942	942	943	943	943	944
8.8	0.944	945	945	946	946	947	947	948	948	949
8.9	0.949	950	950	951	951	952	952	953	953	954
9.0	0.954	955	955	956	956	957	957	958	958	959
9.1	0.959	960	960	960	961	961	962	962	963	963
9.2	0.964	964	965	965	966	966	967	967	968	968
9.3	0.968	969	969	970	970	971	971	972	972	973
9.4	0.973	974	974	975	975	975	976	976	977	977
9.5	0.978	978	979	979	980	980	980	981	981	982
9.6	0.982	983	983	984	984	985	985	985	986	986
9.7	0.987	987	988	988	989	989	989	990	990	991
9.8	0.991	992	992	993	993	993	994	994	995	995
9.9	0.996	996	997	997	997	998	998	999	999	1.000
10.0	1.000									

Squares

	0	1	2	3	4	5	6	7	8	9
1.0	1.00	1.02	1.04	1.06	1.08	1.10	1.12	1.14	1.17	1.19
1.1	1.21	1.23	1.25	1.28	1.30	1.32	1.35	1.37	1.39	1.42
1.2	1.44	1.46	1.49	1.51	1.54	1.56	1.59	1.61	1.64	1.66
1.3	1.69	1.72	1.74	1.77	1.80	1.82	1.85	1.88	1.90	1.93
1.4	1.96	1.99	2.02	2.04	2.07	2.10	2.13	2.16	2.19	2.22
1.5	2.25	2.28	2.31	2.34	2.37	2.40	2.43	2.46	2.50	2.53
1.6	2.56	2.59	2.62	2.66	2.69	2.72	2.76	2.79	2.82	2.86
1.7	2.89	2.92	2.96	2.99	3.03	3.06	3.10	3.13	3.17	3.20
1.8	3.24	3.28	3.31	3.35	3.39	3.42	3.46	3.50	3.53	3.57
1.9	3.61	3.65	3.69	3.72	3.76	3.80	3.84	3.88	3.92	3.96
2.0	4.00	4.04	4.08	4.12	4.16	4.20	4.24	4.28	4.33	4.37
2.1	4.41	4.45	4.49	4.54	4.58	4.62	4.67	4.71	4.75	4.80
2.2	4.84	4.88	4.93	4.97	5.02	5.06	5.11	5.15	5.20	5.24
2.3	5.29	5.34	5.38	5.43	5.48	5.52	5.57	5.62	5.66	5.71
2.4	5.76	5.81	5.86	5.90	5.95	6.00	6.05	6.10	6.15	6.20
2.5	6.25	6.30	6.35	6.40	6.45	6.50	6.55	6.60	6.66	6.71
2.6	6.76	6.81	6.86	6.92	6.97	7.02	7.08	7.13	7.18	7.24
2.7	7.29	7.34	7.40	7.45	7.51	7.56	7.62	7.67	7.73	7.78
2.8	7.84	7.90	7.95	8.01	8.07	8.12	8.18	8.24	8.29	8.35
2.9	8.41	8.47	8.53	8.58	8.64	8.70	8.76	8.82	8.88	8.94
3.0	9.00	9.06	9.12	9.18	9.24	9.30	9.36	9.42	9.49	9.55
3.1	9.61	9.67	9.73	9.80	9.86	9.92	9.99	10.0	10.1	10.2
3.2	10.2	10.3	10.4	10.4	10.5	10.6	10.6	10.7	10.8	10.8
3.3	10.9	11.0	11.0	11.1	11.2	11.2	11.3	11.4	11.4	11.5
3.4	11.6	11.6	11.7	11.8	11.8	11.9	12.0	12.0	12.1	12.2
3.5	12.3	12.3	12.4	12.5	12.5	12.6	12.7	12.7	12.8	12.9
3.6	13.0	13.0	13.1	13.2	13.2	13.3	13.4	13.5	13.5	13.6
3.7	13.7	13.8	13.8	13.9	14.0	14.1	14.1	14.2	14.3	14.4
3.8	14.4	14.5	14.6	14.7	14.7	14.8	14.9	15.0	15.1	15.1
3.9	15.2	15.3	15.4	15.4	15.5	15.6	15.7	15.8	15.8	15.9
4.0	16.0	16.1	16.2	16.2	16.3	16.4	16.5	16.6	16.6	16.7
4.1	16.8	16.9	17.0	17.1	17.1	17.2	17.3	17.4	17.5	17.6
4.2	17.6	17.7	17.8	17.9	18.0	18.1	18.1	18.2	18.3	18.4
4.3	18.5	18.6	18.7	18.7	18.8	18.9	19.0	19.1	19.2	19.3
4.4	19.4	19.4	19.5	19.6	19.7	19.8	19.9	20.0	20.1	20.2
4.5	20.3	20.3	20.4	20.5	20.6	20.7	20.8	20.9	21.0	21.1
4.6	21.2	21.3	21.3	21.4	21.5	21.6	21.7	21.8	21.9	22.0
4.7	22.1	22.2	22.3	22.4	22.5	22.6	22.7	22.8	22.8	22.9
4.8	23.0	23.1	23.2	23.3	23.4	23.5	23.6	23.7	23.8	23.9
4.9	24.0	24.1	24.2	24.3	24.4	24.5	24.6	24.7	24.8	24.9
5.0	25.0	25.1	25.2	25.3	25.4	25.5	25.6	25.7	25.8	25.9
5.1	26.0	26.1	26.2	26.3	26.4	26.5	26.6	26.7	26.8	26.9
5.2	27.0	27.1	27.2	27.4	27.5	27.6	27.7	27.8	27.9	28.0
5.3	28.1	28.2	28.3	28.4	28.5	28.6	28.7	28.8	28.9	29.1
5.4	29.2	29.3	29.4	29.5	29.6	29.7	29.8	29.9	30.0	30.1

	0	1	2	3	4	5	6	7	8	9
5.5	30.3	30.4	30.5	30.6	30.7	30.8	30.9	31.0	31.1	31.2
5.6	31.4	31.5	31.6	31.7	31.8	31.9	32.0	32.1	32.3	32.4
5.7	32.5	32.6	32.7	32.8	32.9	33.1	33.2	33.3	33.4	33.5
5.8	33.6	33.8	33.9	34.0	34.1	34.2	34.3	34.5	34.6	34.7
5.9	34.8	34.9	35.0	35.2	35.3	35.4	35.5	35.6	35.8	35.9
6.0	36.0	36.1	36.2	36.4	36.5	36.6	36.7	36.8	37.0	37.1
6.1	37.2	37.3	37.5	37.6	37.7	37.8	37.9	38.1	38.2	38.3
6.2	38.4	38.6	38.7	38.8	38.9	39.1	39.2	39.3	39.4	39.6
6.3	39.7	39.8	39.9	40.1	40.2	40.3	40.4	40.6	40.7	40.8
6.4	41.0	41.1	41.2	41.3	41.5	41.6	41.7	41.9	42.0	42.1
6.5	42.3	42.4	42.5	42.6	42.8	42.9	43.0	43.2	43.3	43.4
6.6	43.6	43.7	43.8	44.0	44.1	44.2	44.4	44.5	44.6	44.8
6.7	44.9	45.0	45.2	45.3	45.4	45.6	45.7	45.8	46.0	46.1
6.8	46.2	46.4	46.5	46.6	46.8	46.9	47.1	47.2	47.3	47.5
6.9	47.6	47.7	47.9	48.0	48.2	48.3	48.4	48.6	48.7	48.9
7.0	49.0	49.1	49.3	49.4	49.6	49.7	49.8	50.0	50.1	50.3
7.1	50.4	50.6	50.7	50.8	51.0	51.1	51.3	51.4	51.6	51.7
7.2	51.8	52.0	52.1	52.3	52.4	52.6	52.7	52.9	53.0	53.1
7.3	53.3	53.4	53.6	53.7	53.9	54.0	54.2	54.3	54.5	54.6
7.4	54.8	54.9	55.1	55.2	55.4	55.5	55.7	55.8	56.0	56.1
7.5	56.3	56.4	56.6	56.7	56.9	57.0	57.2	57.3	57.5	57.6
7.6	57.8	57.9	58.1	58.2	58.4	58.5	58.7	58.8	59.0	59.1
7.7	59.3	59.4	59.6	59.8	59.9	60.1	60.2	60.4	60.5	60.7
7.8	60.8	61.0	61.2	61.3	61.5	61.6	61.8	61.9	62.1	62.3
7.9	62.4	62.6	62.7	62.9	63.0	63.2	63.4	63.5	63.7	63.8
8.0	64.0	64.2	64.3	64.5	64.6	64.8	65.0	65.1	65.3	65.4
8.1	65.6	65.8	65.9	66.1	66.3	66.4	66.6	66.7	66.9	67.1
8.2	67.2	67.4	67.6	67.7	67.9	68.1	68.2	68.4	68.6	68.7
8.3	68.9	69.1	69.2	69.4	69.6	69.7	69.9	70.1	70.2	70.4
8.4	70.6	70.7	70.9	71.1	71.2	71.4	71.6	71.7	71.9	72.1
8.5	72.3	72.4	72.6	72.8	72.9	73.1	73.3	73.4	73.6	73.8
8.6	74.0	74.1	74.3	74.5	74.6	74.8	75.0	75.2	75.3	75.5
8.7	75.7	75.9	76.0	76.2	76.4	76.6	76.7	76.9	77.1	77.3
8.8	77.4	77.6	77.8	78.0	78.1	78.3	78.5	78.7	78.9	79.0
8.9	79.2	79.4	79.6	79.7	79.9	80.1	80.3	80.5	80.6	80.8
9.0	81.0	81.2	81.4	81.5	81.7	81.9	82.1	82.3	82.4	82.6
9.1	82.8	83.0	83.2	83.4	83.5	83.7	83.9	84.1	84.3	84.5
9.2	84.6	84.8	85.0	85.2	85.4	85.6	85.7	85.9	86.1	86.3
9.3	86.5	86.7	86.9	87.0	87.2	87.4	87.6	87.8	88.0	88.2
9.4	88.4	88.5	88.7	88.9	89.1	89.3	89.5	89.7	89.9	90.1
9.5	90.3	90.4	90.6	90.8	91.0	91.2	91.4	91.6	91.8	92.0
9.6	92.2	92.4	92.5	92.7	92.9	93.1	93.3	93.5	93.7	93.9
9.7	94.1	94.3	94.5	94.7	94.9	95.1	95.3	95.5	95.6	95.8
9.8	96.0	96.2	96.4	96.6	96.8	97.0	97.2	97.4	97.6	97.8
9.9	98.0	98.2	98.4	98.6	98.8	99.0	99.2	99.4	99.6	99.8
10.0	100									

Square Roots of Numbers from 1 to 99

	0	1	2	3	4	5	6	7	8	9
1.0	1.00	1.00	1.01	1.01	1.02	1.02	1.03	1.03	1.04	1.04
1.1	1.05	1.05	1.06	1.06	1.07	1.07	1.08	1.08	1.09	1.09
1.2	1.10	1.10	1.10	1.11	1.11	1.12	1.12	1.13	1.13	1.14
1.3	1.14	1.14	1.15	1.15	1.16	1.16	1.17	1.17	1.17	1.18
1.4	1.18	1.19	1.19	1.20	1.20	1.20	1.21	1.21	1.22	1.22
1.5	1.22	1.23	1.23	1.24	1.24	1.24	1.25	1.25	1.26	1.26
1.6	1.26	1.27	1.27	1.28	1.28	1.28	1.29	1.29	1.30	1.30
1.7	1.30	1.31	1.31	1.32	1.32	1.32	1.33	1.33	1.33	1.34
1.8	1.34	1.35	1.35	1.35	1.36	1.36	1.36	1.37	1.37	1.37
1.9	1.38	1.38	1.39	1.39	1.39	1.40	1.40	1.40	1.41	1.41
2.0	1.41	1.42	1.42	1.42	1.43	1.43	1.44	1.44	1.44	1.45
2.1	1.45	1.45	1.46	1.46	1.46	1.47	1.47	1.47	1.48	1.48
2.2	1.48	1.49	1.49	1.49	1.50	1.50	1.50	1.51	1.51	1.51
2.3	1.52	1.52	1.52	1.53	1.53	1.53	1.54	1.54	1.54	1.55
2.4	1.55	1.55	1.56	1.56	1.56	1.57	1.57	1.57	1.57	1.58
2.5	1.58	1.58	1.59	1.59	1.59	1.60	1.60	1.60	1.61	1.61
2.6	1.61	1.62	1.62	1.62	1.62	1.63	1.63	1.63	1.64	1.64
2.7	1.64	1.65	1.65	1.65	1.66	1.66	1.66	1.66	1.67	1.67
2.8	1.67	1.68	1.68	1.68	1.69	1.69	1.69	1.69	1.70	1.70
2.9	1.70	1.71	1.71	1.71	1.71	1.72	1.72	1.72	1.73	1.73
3.0	1.73	1.73	1.74	1.74	1.74	1.75	1.75	1.75	1.75	1.76
3.1	1.76	1.76	1.77	1.77	1.77	1.77	1.78	1.78	1.78	1.79
3.2	1.79	1.79	1.79	1.80	1.80	1.80	1.81	1.81	1.81	1.81
3.3	1.82	1.82	1.82	1.82	1.83	1.83	1.83	1.84	1.84	1.84
3.4	1.84	1.85	1.85	1.85	1.85	1.86	1.86	1.86	1.87	1.87
3.5	1.87	1.87	1.88	1.88	1.88	1.88	1.89	1.89	1.89	1.89
3.6	1.90	1.90	1.90	1.91	1.91	1.91	1.91	1.92	1.92	1.92
3.7	1.92	1.93	1.93	1.93	1.93	1.94	1.94	1.94	1.94	1.95
3.8	1.95	1.95	1.95	1.96	1.96	1.96	1.96	1.97	1.97	1.97
3.9	1.97	1.98	1.98	1.98	1.98	1.99	1.99	1.99	1.99	2.00
4.0	2.00	2.00	2.00	2.01	2.01	2.01	2.01	2.02	2.02	2.02
4.1	2.02	2.03	2.03	2.03	2.03	2.04	2.04	2.04	2.04	2.05
4.2	2.05	2.05	2.05	2.06	2.06	2.06	2.06	2.07	2.07	2.07
4.3	2.07	2.08	2.08	2.08	2.08	2.09	2.09	2.09	2.09	2.10
4.4	2.10	2.10	2.10	2.10	2.11	2.11	2.11	2.11	2.12	2.12
4.5	2.12	2.12	2.13	2.13	2.13	2.13	2.14	2.14	2.14	2.14
4.6	2.14	2.15	2.15	2.15	2.15	2.16	2.16	2.16	2.16	2.17
4.7	2.17	2.17	2.17	2.17	2.18	2.18	2.18	2.18	2.19	2.19
4.8	2.19	2.19	2.20	2.20	2.20	2.20	2.20	2.21	2.21	2.21
4.9	2.21	2.22	2.22	2.22	2.22	2.22	2.23	2.23	2.23	2.23
5.0	2.24	2.24	2.24	2.24	2.24	2.25	2.25	2.25	2.25	2.26
5.1	2.26	2.26	2.26	2.26	2.27	2.27	2.27	2.27	2.28	2.28
5.2	2.28	2.28	2.28	2.29	2.29	2.29	2.29	2.30	2.30	2.30
5.3	2.30	2.30	2.31	2.31	2.31	2.31	2.32	2.32	2.32	2.32
5.4	2.32	2.33	2.33	2.33	2.33	2.33	2.34	2.34	2.34	2.34
5.5	2.35	2.35	2.35	2.35	2.36	2.36	2.36	2.36	2.36	2.36
5.6	2.37	2.37	2.37	2.37	2.37	2.38	2.38	2.38	2.38	2.39
5.7	2.39	2.39	2.39	2.39	2.40	2.40	2.40	2.40	2.40	2.41
5.8	2.41	2.41	2.41	2.41	2.42	2.42	2.42	2.42	2.42	2.43
5.9	2.43	2.43	2.43	2.44	2.44	2.44	2.44	2.44	2.45	2.45
6.0	2.45	2.45	2.45	2.46	2.46	2.46	2.46	2.46	2.47	2.47
6.1	2.47	2.47	2.47	2.48	2.48	2.48	2.48	2.48	2.49	2.49
6.2	2.49	2.49	2.49	2.50	2.50	2.50	2.50	2.50	2.51	2.51
6.3	2.51	2.51	2.51	2.52	2.52	2.52	2.52	2.52	2.53	2.53
6.4	2.53	2.53	2.53	2.54	2.54	2.54	2.54	2.54	2.55	2.55
6.5	2.55	2.55	2.55	2.56	2.56	2.56	2.56	2.56	2.57	2.57
6.6	2.57	2.57	2.57	2.57	2.58	2.58	2.58	2.58	2.58	2.59
6.7	2.59	2.59	2.59	2.59	2.60	2.60	2.60	2.60	2.60	2.61
6.8	2.61	2.61	2.61	2.61	2.62	2.62	2.62	2.62	2.62	2.62
6.9	2.63	2.63	2.63	2.63	2.63	2.64	2.64	2.64	2.64	2.64

Square Roots of Numbers from 1 to 99

	0	1	2	3	4	5	6	7	8	9
7.0	2.65	2.65	2.65	2.65	2.65	2.66	2.66	2.66	2.66	2.66
7.1	2.66	2.67	2.67	2.67	2.67	2.67	2.68	2.68	2.68	2.68
7.2	2.68	2.69	2.69	2.69	2.69	2.69	2.69	2.70	2.70	2.70
7.3	2.70	2.70	2.71	2.71	2.71	2.71	2.71	2.71	2.72	2.72
7.4	2.72	2.72	2.72	2.73	2.73	2.73	2.73	2.73	2.73	2.74
7.5	2.74	2.74	2.74	2.74	2.75	2.75	2.75	2.75	2.75	2.75
7.6	2.76	2.76	2.76	2.76	2.76	2.77	2.77	2.77	2.77	2.77
7.7	2.77	2.78	2.78	2.78	2.78	2.78	2.79	2.79	2.79	2.79
7.8	2.79	2.79	2.80	2.80	2.80	2.80	2.80	2.81	2.81	2.81
7.9	2.81	2.81	2.81	2.82	2.82	2.82	2.82	2.82	2.82	2.83
8.0	2.83	2.83	2.83	2.83	2.84	2.84	2.84	2.84	2.84	2.84
8.1	2.85	2.85	2.85	2.85	2.85	2.85	2.86	2.86	2.86	2.86
8.2	2.86	2.87	2.87	2.87	2.87	2.87	2.87	2.88	2.88	2.88
8.3	2.88	2.88	2.88	2.89	2.89	2.89	2.89	2.89	2.89	2.90
8.4	2.90	2.90	2.90	2.90	2.91	2.91	2.91	2.91	2.91	2.91
8.5	2.92	2.92	2.92	2.92	2.92	2.92	2.93	2.93	2.93	2.93
8.6	2.93	2.93	2.94	2.94	2.94	2.94	2.94	2.94	2.95	2.95
8.7	2.95	2.95	2.95	2.95	2.96	2.96	2.96	2.96	2.96	2.96
8.8	2.97	2.97	2.97	2.97	2.97	2.97	2.98	2.98	2.98	2.98
8.9	2.98	2.98	2.99	2.99	2.99	2.99	2.99	2.99	3.00	3.00
9.0	3.00	3.00	3.00	3.00	3.01	3.01	3.01	3.01	3.01	3.01
9.1	3.02	3.02	3.02	3.02	3.02	3.02	3.03	3.03	3.03	3.03
9.2	3.03	3.03	3.04	3.04	3.04	3.04	3.04	3.04	3.05	3.05
9.3	3.05	3.05	3.05	3.05	3.06	3.06	3.06	3.06	3.06	3.06
9.4	3.07	3.07	3.07	3.07	3.07	3.07	3.08	3.08	3.08	3.08
9.5	3.08	3.08	3.09	3.09	3.09	3.09	3.09	3.09	3.10	3.10
9.6	3.10	3.10	3.10	3.10	3.10	3.11	3.11	3.11	3.11	3.11
9.7	3.11	3.12	3.12	3.12	3.12	3.12	3.12	3.13	3.13	3.13
9.8	3.13	3.13	3.13	3.14	3.14	3.14	3.14	3.14	3.14	3.14
9.9	3.15	3.15	3.15	3.15	3.15	3.15	3.16	3.16	3.16	3.16
10	3.16	3.18	3.19	3.21	3.22	3.24	3.26	3.27	3.29	3.30
11	3.32	3.33	3.35	3.36	3.38	3.39	3.41	3.42	3.44	3.45
12	3.46	3.48	3.49	3.51	3.52	3.54	3.55	3.56	3.58	3.59
13	3.61	3.62	3.63	3.65	3.66	3.67	3.69	3.70	3.71	3.73
14	3.74	3.75	3.77	3.78	3.79	3.81	3.82	3.83	3.85	3.86
15	3.87	3.89	3.90	3.91	3.92	3.94	3.95	3.96	3.97	3.99
16	4.00	4.01	4.02	4.04	4.05	4.06	4.07	4.09	4.10	4.11
17	4.12	4.14	4.15	4.16	4.17	4.18	4.20	4.21	4.22	4.23
18	4.24	4.25	4.27	4.28	4.29	4.30	4.31	4.32	4.34	4.35
19	4.36	4.37	4.38	4.39	4.40	4.42	4.43	4.44	4.45	4.46
20	4.47	4.48	4.49	4.51	4.52	4.53	4.54	4.55	4.56	4.57
21	4.58	4.59	4.60	4.62	4.63	4.64	4.65	4.66	4.67	4.68
22	4.69	4.70	4.71	4.72	4.73	4.74	4.75	4.76	4.77	4.79
23	4.80	4.81	4.82	4.83	4.84	4.85	4.86	4.87	4.88	4.89
24	4.90	4.91	4.92	4.93	4.94	4.95	4.96	4.97	4.98	4.99
25	5.00	5.01	5.02	5.03	5.04	5.05	5.06	5.07	5.08	5.09
26	5.10	5.11	5.12	5.13	5.14	5.15	5.16	5.17	5.18	5.19
27	5.20	5.21	5.22	5.22	5.23	5.24	5.25	5.26	5.27	5.28
28	5.29	5.30	5.31	5.32	5.33	5.34	5.35	5.36	5.37	5.38
29	5.39	5.39	5.40	5.41	5.42	5.43	5.44	5.45	5.46	5.47
30	5.48	5.49	5.50	5.50	5.51	5.52	5.53	5.54	5.55	5.56
31	5.57	5.58	5.59	5.59	5.60	5.61	5.62	5.63	5.64	5.65
32	5.66	5.67	5.67	5.68	5.69	5.70	5.71	5.72	5.73	5.74
33	5.74	5.75	5.76	5.77	5.78	5.79	5.80	5.81	5.81	5.82
34	5.83	5.84	5.85	5.86	5.87	5.87	5.88	5.89	5.90	5.91
35	5.92	5.92	5.93	5.94	5.95	5.96	5.97	5.97	5.98	5.99
36	6.00	6.01	6.02	6.02	6.03	6.04	6.05	6.06	6.07	6.07
37	6.08	6.09	6.10	6.11	6.12	6.12	6.13	6.14	6.15	6.16
38	6.16	6.17	6.18	6.19	6.20	6.20	6.21	6.22	6.23	6.24
39	6.24	6.25	6.26	6.27	6.28	6.28	6.29	6.30	6.31	6.32

	0	1	2	3	4	5	6	7	8	9
40	6.32	6.33	6.34	6.35	6.36	6.36	6.37	6.38	6.39	6.40
41	6.40	6.41	6.42	6.43	6.43	6.44	6.45	6.46	6.47	6.47
42	6.48	6.49	6.50	6.50	6.51	6.52	6.53	6.53	6.54	6.55
43	6.56	6.57	6.57	6.58	6.59	6.60	6.60	6.61	6.62	6.63
44	6.63	6.64	6.65	6.66	6.66	6.67	6.68	6.69	6.69	6.70
45	6.71	6.72	6.72	6.73	6.74	6.75	6.75	6.76	6.77	6.77
46	6.78	6.79	6.80	6.80	6.81	6.82	6.83	6.83	6.84	6.85
47	6.86	6.86	6.87	6.88	6.88	6.89	6.90	6.91	6.91	6.92
48	6.93	6.94	6.94	6.95	6.96	6.96	6.97	6.98	6.99	6.99
49	7.00	7.01	7.01	7.02	7.03	7.04	7.04	7.05	7.06	7.06
50	7.07	7.08	7.09	7.09	7.10	7.11	7.11	7.12	7.13	7.13
51	7.14	7.15	7.16	7.16	7.17	7.18	7.18	7.19	7.20	7.20
52	7.21	7.22	7.22	7.23	7.24	7.25	7.25	7.26	7.27	7.27
53	7.28	7.29	7.29	7.30	7.31	7.31	7.32	7.33	7.33	7.34
54	7.35	7.36	7.36	7.37	7.38	7.38	7.39	7.40	7.40	7.41
55	7.42	7.42	7.43	7.44	7.44	7.45	7.46	7.46	7.47	7.48
56	7.48	7.49	7.50	7.50	7.51	7.52	7.52	7.53	7.54	7.54
57	7.55	7.56	7.56	7.57	7.58	7.58	7.59	7.60	7.60	7.61
58	7.62	7.62	7.63	7.64	7.64	7.65	7.66	7.66	7.67	7.67
59	7.68	7.69	7.69	7.70	7.71	7.71	7.72	7.73	7.73	7.74
60	7.75	7.75	7.76	7.77	7.77	7.78	7.78	7.79	7.80	7.80
61	7.81	7.82	7.82	7.83	7.84	7.84	7.85	7.85	7.86	7.87
62	7.87	7.88	7.89	7.89	7.90	7.91	7.91	7.92	7.92	7.93
63	7.94	7.94	7.95	7.96	7.96	7.97	7.97	7.98	7.99	7.99
64	8.00	8.01	8.01	8.02	8.02	8.03	8.04	8.04	8.05	8.06
65	8.06	8.07	8.07	8.08	8.09	8.09	8.10	8.11	8.11	8.12
66	8.12	8.13	8.14	8.14	8.15	8.15	8.16	8.17	8.17	8.18
67	8.19	8.19	8.20	8.20	8.21	8.22	8.22	8.23	8.23	8.24
68	8.25	8.25	8.26	8.26	8.27	8.28	8.28	8.29	8.29	8.30
69	8.31	8.31	8.32	8.32	8.33	8.34	8.34	8.35	8.35	8.36
70	8.37	8.37	8.38	8.38	8.39	8.40	8.40	8.41	8.41	8.42
71	8.43	8.43	8.44	8.44	8.45	8.46	8.46	8.47	8.47	8.48
72	8.49	8.49	8.50	8.50	8.51	8.51	8.52	8.53	8.53	8.54
73	8.54	8.55	8.56	8.56	8.57	8.57	8.58	8.58	8.59	8.60
74	8.60	8.61	8.61	8.62	8.63	8.63	8.64	8.64	8.65	8.65
75	8.66	8.67	8.67	8.68	8.68	8.69	8.69	8.70	8.71	8.71
76	8.72	8.72	8.73	8.73	8.74	8.75	8.75	8.76	8.76	8.77
77	8.77	8.78	8.79	8.79	8.80	8.80	8.81	8.81	8.82	8.83
78	8.83	8.84	8.84	8.85	8.85	8.86	8.87	8.87	8.88	8.88
79	8.89	8.89	8.90	8.91	8.91	8.92	8.92	8.93	8.93	8.94
80	8.94	8.95	8.96	8.96	8.97	8.97	8.98	8.98	8.99	8.99
81	9.00	9.01	9.01	9.02	9.02	9.03	9.03	9.04	9.04	9.05
82	9.06	9.06	9.07	9.07	9.08	9.08	9.09	9.09	9.10	9.10
83	9.11	9.12	9.12	9.13	9.13	9.14	9.14	9.15	9.15	9.16
84	9.17	9.17	9.18	9.18	9.19	9.19	9.20	9.20	9.21	9.21
85	9.22	9.22	9.23	9.24	9.24	9.25	9.25	9.26	9.26	9.27
86	9.27	9.28	9.28	9.29	9.30	9.30	9.31	9.31	9.32	9.32
87	9.33	9.33	9.34	9.34	9.35	9.35	9.36	9.36	9.37	9.38
88	9.38	9.39	9.39	9.40	9.40	9.41	9.41	9.42	9.42	9.43
89	9.43	9.44	9.44	9.45	9.46	9.46	9.47	9.47	9.48	9.48
90	9.49	9.49	9.50	9.50	9.51	9.51	9.52	9.52	9.53	9.53
91	9.54	9.54	9.55	9.56	9.56	9.57	9.57	9.58	9.58	9.59
92	9.59	9.60	9.60	9.61	9.61	9.62	9.62	9.63	9.63	9.64
93	9.64	9.65	9.65	9.66	9.66	9.67	9.67	9.68	9.69	9.69
94	9.70	9.70	9.71	9.71	9.72	9.72	9.73	9.73	9.74	9.74
95	9.75	9.75	9.76	9.76	9.77	9.77	9.78	9.78	9.79	9.79
96	9.80	9.80	9.81	9.81	9.82	9.82	9.83	9.83	9.84	9.84
97	9.85	9.85	9.86	9.86	9.87	9.87	9.88	9.88	9.89	9.89
98	9.90	9.90	9.91	9.91	9.92	9.92	9.93	9.93	9.94	9.94
99	9.95	9.95	9.96	9.96	9.97	9.97	9.98	9.98	9.99	9.99